RAISING ENGINEERS

RAISING ENGINEERS

A Founder's
Guide to Building a
**High-Performing
Engineering
Team**

DAVID D. DETTMER

LIONCREST
PUBLISHING

Raising Engineers

A Founder's Guide to Building a High-Performing Engineering Team

ISBN 978-1-5445-2764-2 *Hardcover*

978-1-5445-2762-8 *Paperback*

978-1-5445-2763-5 *Ebook*

To my parents for raising me,

to my wife, Robyn, for all her support,

and to my daughter, Bailey,

the engineer I raised.

CONTENTS

INTRODUCTION

Over the years, I've had hundreds of conversations with tech startup founders that sound something like this:

"I've recently raised some capital and now I need to hire engineers and build a team. How do I attract the right engineers and how do I ensure the team is productive?"

One of the most personally memorable of these conversations was in Austin, Texas in 2014. I met a founder in downtown Austin at a burger joint called Wholly Cow. He had been going around the country looking for other possible locations to expand his engineering team.

"We've recently raised $8 million. I've got five engineers and the funding to hire more, but growing the team has been a struggle," the founder told me. "I'm currently flying around the United States meeting with engineering leaders trying to figure out if I need to build a team outside of San Francisco."

I recognized his pain right away because I'd heard it from hundreds of founders before. Even with a sizable amount of cash available for hiring, putting together a software engineering team is not easy.

I started with the question I ask all founders: "What does your company do?"

His answer would change the next seven years of my life, all the way up to the present day.

"We create work-from-home jobs," he said.

I knew instantly I wanted to be a part of that. For one thing, this founder already had in place the very first thing I tell founders they need to recruit engineers. He had a clear mission. He could've answered, "We sell speech-to-text services" or "We caption movies," but he knew that was only what his company sold. Instead, he told me why his company existed.

When I heard it, I could instantly feel its power. Work-from-home jobs change people's lives by allowing them to work where, when, and how they want.

Soon after that, I became the leader of the engineering team for this founder's company, Rev.com. I'm still there and the mission keeps my motivational fire lit to this day. We've built a team that's grown from a handful of engineers to more than a hundred, with about one-third of those engineers being MIT graduates. I'm pleased to say, this team has powered amazing growth at Rev and the company has gone from startup to success story.

All the key concepts I've learned to grow engineering teams at Rev and several other companies are in this book. And if you're reading this, I'm guessing you know how that founder felt. You already know exactly how precious early investment dollars are. You also grasp that a high-performing engineering team is crucial to the success of a startup and are painfully aware of the flipside: a poor team can wreck a company.

The founders I talk to also typically have the following in common:

- ▸ They're savvy and smart, and their companies have enough promise to have secured serious investment dollars.

- ▸ They have anywhere from one to five engineers on board already, and now they need to hire anywhere from five to twenty more (at least to start).

- ▸ This is their first time building an engineering team and they're looking for advice they can trust.

If all this resonates with you, read on. You'll learn both the high-level concepts plus the concrete strategies I've used to build high-performing engineering teams.

MY BACKGROUND IN STARTUPS

The super-brief version of my résumé would read something like this: 1995 MIT grad, then a couple of years as a software engineer, followed by leadership roles for more than two decades, including several VP of Engineering positions. But those are titles and roles, and what you care about is this: can you have confidence that what I tell you will help you build a high-performing engineering team?

I believe so. My twenty-five-plus-year career has been exclusively with startups.

One of the most formative experiences happened early in my career in the late 1990s at the now-legendary tech company Trilogy. Alumni from Trilogy have gone on to become founders at Carta, Capital Factory, Mozilla, and other successful companies.

Trilogy was one of the first tech companies to leave California for Austin. I don't think it's an exaggeration to say that Austin is a tech mecca today in large part because of Trilogy.

What made it special, however, was not that it was a pioneer in moving to Austin. The difference was in its commitment to building a special culture. Trilogy invested millions in hiring thousands of people who were the right fit for their organization.

I remember being flown to Austin for an interview. They put me on a boat and let me interact with their team. It was eye-opening to see how much people loved working there. It was infectious, and I recall thinking that this was definitely not the standard job hiring process, that the "interviews" were actually fun and gave a true window on what it was like to work there. (And this was the late 1990s, before other companies adopted Trilogy-like hiring strategies.)

Once I came on board, I loved Trilogy even more. No one told me I had to stay up all night and code, but I did it anyway. When you love your work and company that much, you're happy to do it.

It was a special time, and no doubt some of it was because it was exciting to be young and at a company that was making such a big impact. But it was more than that to me. It also laid the foundation for my journey to discover how to create similar magic at other companies. In many ways, the heart of the ideas you'll find in this book grew out of that one experience.

Of course, I've had many additional experiences since then. I've been in the trenches. I understand startup tech companies, and

I understand the strategies and methods to help find and hire engineers who are the right fit. I've hired hundreds of engineers personally and advised on the hiring of thousands.

In my work with successful startups, I've met many of the top-name tech investors. So now after years and years of doing this, my core network includes fifty or so high-end investors who regularly put money into tech startups, and they repeatedly send founders to me for advice on building engineering teams. Through those connections, my view on raising engineering teams has expanded well beyond the startups I've been directly involved with.

Having experience specific to startups matters. There's a somewhat common misunderstanding among founders that the best advice on building engineering teams would come from behemoths like Netflix or Facebook. It's an understandable misconception because obviously these companies are extraordinarily successful. Why wouldn't you want to follow the strategies of the biggest and most successful companies?

Setting aside whether you could find someone at one of these companies to advise you, the truth is, it wouldn't be relevant to where you are now anyway. Huge tech companies are at a completely different stage of their life cycle, and their hiring and recruiting is a different beast altogether.

You need someone who can guide you from a small handful of engineers—maybe even only one—to the next five. And then the next five to ten after that, and so on. My advice is directly relevant to you because it's been my life's work to create high-performing engineering teams from scratch—and I've done it again and again. Of course, I've also made many mistakes along the way. I'll also tell you what I've learned from my missteps so you can avoid them.

As you go through the book, I'll share specific examples from my experience, particularly my most recent startup adventure at Rev.

I've also invested in several startups over the years, and this has given me even wider experience, connections, and contacts. Within the community of tech startups, particularly in Austin, this has snowballed into a reputation for building excellent engineering teams.

After talking with and advising literally hundreds of tech startup founders, I have a good handle on what works (and what doesn't) for building strong teams. I'm sharing it all in this book.

WHAT THIS BOOK CAN DO FOR YOU

When you're finished reading this book, you'll know how to recruit and motivate a high-performing engineering team, and how to optimize your company culture and processes for maximum success.

As you go through this book, you'll notice certain themes recur in many of the chapters. Two of the most important are implied in the title, *Raising Engineers*.

One of those ideas is that recruiting and hiring good engineers is only the beginning. Sometimes founders think that building a great team is mostly about hiring. That's not enough. You also have to "raise" your team the right way. You need to find ways to give your engineers purpose, you need to give them some autonomy, and you've got to give them opportunities to master their craft.

Hiring does matter, of course, but if you don't create a work environment where your engineering team is motivated, and if you don't optimize the processes they work within, the best recruitment in the world won't save you.

The other idea embedded in the title is the idea of caring. When parents do a good job of raising a family, it's because they truly care and they show it. A thread that runs through this entire book is that you need to genuinely care about your team as people and look out for their best interests, too—not just your company's.

Many of the founders I talk to are frustrated because they think team members aren't dedicated enough. But these same founders are failing to ask themselves this: are they showing the same level of dedication to the individuals who are on their team?

One of the cornerstone concepts of this book is that a founder needs to show true caring for the engineers who work for them and a willingness to be vulnerable and honest about the company and its mission.

THE STRUCTURE OF THE BOOK

Here's a quick map to this book to help you get the most out of it.

The first section of the book is on recruiting. You'll learn that having a clear mission for your company will be crucial for attracting the right engineers, the ones who are the right match for your particular company. You'll also acquire solid strategies for interviewing and "closing"—getting the engineers you want on your team to say yes.

After recruiting, the next section is all about creating a company culture that is both motivating and accountable. The concepts of purpose, autonomy, and mastery will be fundamental to establishing a healthy work environment. These are important ideas outlined in Daniel Pink's book *Drive*, and I'll show you how to adapt and implement them specifically for engineering teams.

The last major section of the book is about optimizing your people and your processes. Achieving velocity, and how to define that, are the focus. I'll show you what's important to pay attention to when getting your team to go faster, and how to avoid getting bogged down with tasks that don't matter.

The final chapter brings it all together and reminds you that it's not enough to care about the company yourself; you've got to find ways to make that caring real for your team.

Of course, that summary of the book probably sounds like neatly divided sections and a step-by-step recipe for success. First, you recruit. Then you build a culture of motivation. Then you optimize that culture and its processes. You finish one thing and then move to the next, right? Of course not.

You'll need to be doing all these things at the same time. The good news is that it's not as hard as it sounds at first. The concepts I advocate all interlock, and each element reinforces the others.

When you recruit engineers who are the right fit for your company, motivating and optimizing will be easier. When you optimize using the principles in this book, you'll discover that reinforces motivation. Everything begins to work together to increase velocity and get your startup over the tipping point where success breeds more success and you achieve the sustained growth you need.

There's one more point that will help you understand how to get the most out of this book, and that's to understand what this book isn't. This is not a rigid instruction manual. It does have practical strategies, and it is systematic to some degree, but it's not an insert-tab-A-into-slot-B kind of book.

Although I can give you the concepts you'll need to succeed, there's never a neat, one-size-fits-all solution in the world of startups. The companies and the work we do are too creative and varied for cookie-cutter answers.

That's just as well anyway. Tech startup founders are not give-me-commands kind of people in my experience. They're too smart and independent for that. In my conversations with them, they want models that work and some practical advice, and then they'll run with it. The book is intended to work the same way.

Building a high-performing team starts with hiring, so we'll tackle recruiting first. You might think that means diving into details like how to best use LinkedIn or deciding on a compensation and benefits range. We will eventually cover those details, but that's not where you start.

The first step is figuring out and expressing your company mission in a way that will make it a magnet for engineers who align with your purpose. That's where we'll begin the first chapter.

Chapter 1

ATTRACTING ENGINEERS

Some founders are initially a bit shocked by how hard it can be to attract software engineers to their team.

I remember consulting with a founder who had previously worked at Apple. He told me that every night he'd sit at his desk sending out message after message to try to get people interested in working for his company. He was getting very little response.

"At Apple, everyone just wanted to come work for us," he said. He was finding out just how different attracting engineers is at a startup. I've also met founders who assume that engineers will automatically be excited to join their newly funded startup with a unique offering. Of course, the truth is that attracting engineers is hard, no matter what your previous experience is or how special your startup may be.

The good news is that there are effective ways to do it. Founders just aren't always sure where to begin.

When I ask founders about how they will start hiring more engineers, I typically hear things like '"create an enticing job description" or "reach out to my network" or "just hire a good recruiting agency."

Those answers aren't wrong exactly, but they are the wrong place to start. Those are tactics, and those come later. There's a deeper and more strategic need that must come first if you want to truly connect with the right people for your company.

Great recruiting begins with getting clarity on three things:

- ► The mission of your company

- ► Creating a pitch around what engineers care about

▸ Why you personally want to build
  this company

Get these three right, and you'll have a powerful foundation for attracting engineers who are the right match for your company. The last part of that sentence is especially important. You're not only looking for the best software engineers. You're seeking the engineers who are the best fit for *your* startup.

Let's take a deeper dive into the three keys to finding the right engineers.

RECRUITING =
AN INVITATION TO CONTRIBUTE
TO YOUR COMPANY MISSION

Before you do any recruiting, can you state simply and clearly what your company does? Notice I didn't say what it sells, but what it does. Let me illustrate the difference with an example from my own experience.

As head of engineering at Rev, I've had literally thousands of conversations with prospective engineering candidates. When I ask those candidates what it is they think we do at Rev, they almost inevitably say something that boils down to "you sell transcription and other speech-to-text services."

My response is always the same: "You're right. But that's not why we exist."

Next, I share what we really do at Rev: we create work-from-home jobs. That's our mission. Then I go deeper and explain what that means.

"We change people's lives by allowing them to work where, when, and how they want. You hear every day how companies move to the next city, state, or country or have just shut down permanently and not everyone is as fortunate as you and me and can just work from anywhere or move. They have families, mortgages, and responsibilities, and we create opportunities for these people. Because of Rev, there are thousands of people who can pay that mortgage, put food on the table, or just have some extra cash for the weekend. We change lives."

I'm trying to reach them at a gut level and see how they respond. It's like I'm extending them an invitation to share in the mission of Rev, and that comes before we talk about the actual job and its details. If our mission excites them, then we can go through the rest of the recruitment process to make sure everything else is a match. But the mission has to be the foundation and provide the necessary spark.

Compare this to what would happen if I told engineers, "We sell speech-to-text services." Unless they're a speech scientist, it might come across like any other tech job and would probably do little to distinguish it from other options they may have. On the other hand, letting them in on our true mission—to change lives by creating work-from-home jobs—creates connection and excitement.

Does this mission and pitch connect with every candidate I talk to? Of course not. And that's the point.

Whenever I'm in a conversation with a prospective engineer, I'm watching for signs that the mission generates some spark of interest. However, even if I'm not getting those signals right off the bat, I don't immediately rule them out.

I'll continue to bring up the mission at different points of the conversation. I'm probing to see if they do connect with it and I just haven't found out why yet.

If by the end of the conversation I like a candidate's skills and potential, but I still don't feel any excitement about this mission, I ask two questions: "Are you interested in moving forward with this process? If so, why?"

What I hope to hear is one of three things: that they do connect with the mission, that they're excited about the technical challenges we're tackling, or that they're excited about how the engineering team operates.

On the other hand, some candidates will focus their answers on negative circumstances in their current job. If they're simply looking for an escape valve and don't seem to have any genuinely positive reasons to work for your company, hiring them will likely end badly.

Although ideally, all your engineers will be excited about your mission, there may be a handful who connect with your company for other positive reasons and these can be good hires, too.

That said, alignment with mission needs be a key criterion in deciding whom to hire in most cases. As a founder of a startup in early-growth mode, the first twenty engineers you bring on are crucial for going to the next level. These hires need to be aligned with you and buy into what you do at a very deep level. By recruiting people who are enthusiastic about your mission and want to help you achieve your goal, you'll get motivated engineers on board. These motivated engineers are the people who will fuel your company's growth.

This is why great recruitment all starts with figuring out what you do and putting it in one succinct mission sentence. A good rule of thumb is this: your mission statement should be eight words or less and extremely easy to understand. As in, super clear and immediately graspable to pretty much everyone.

I should also note that you shouldn't confuse the mission statement with an elevator pitch. Most tech founders have an elevator pitch that they use to raise money, which is perfect for that purpose. A mission is something shorter and more visionary, the tagline that sums up the core motivation behind everything your company does.

Some startups already have a terrific mission statement. If that's you, go ahead and skip ahead to the next section in this chapter. But if you only know what you sell but not what you do, here are three great questions to ask yourself to find your mission:

▸ What's going to change in the world because of your company's success?

▸ How are people's lives going to change because of your company?

▸ Why is what your company does important?

Here are a few best practices to keep in mind when crafting a mission statement.

Do It Together

This should most definitely be a collaborative process. Assuming you have a small core group already in place, include them in this process. Your mission should connect your entire team and get everyone fired up.

Take the Time

Building on the previous point, demonstrate the importance of this mission by setting aside significant time to talk it through.

There's no set process for finding your mission, but here's one way to go about it.

As founder, first answer the three questions yourself as a way to get started. Then bring your group together and share those answers and work together on the mission. Take as much time as needed; meeting multiple times about it is not too much. Remember, this will not be just used in hiring; it also needs to inspire your team for a long period of time.

On the other hand, don't stress too much thinking that the mission statement "has to last forever." It does need to be powerful and have staying power, but companies evolve, so don't put too much pressure on yourself to create something that will never change.

Test It

The last step is to put it out there and test your message. You don't have to wait to test it on prospective hires. When the opportunity arises in conversations, state your mission and see if it connects. You'll quickly discover if it's falling flat or if people are sparked by it. In some cases, you may have the mission right, but you might just need to find better words to express it.

Once you're crystal clear on the company mission and have honed it into a succinct sentence, it's now time to create your pitch.

CREATING YOUR PITCH

The focus so far has been on the company mission and your own personal motivations. Now we need to reverse things and think about what this looks like from an engineer's point of view. You'll want to craft your recruitment pitch from their standpoint. Your recruiting pitch will be "delivered" during an organic conversation, so it will be a different presentation style than some of the other pitches you've done, but the principles are the same.

As a founder, I have little doubt that creating a pitch is in your DNA. The reason you're likely reading this book is that you made a pitch for capital and were successful at it and now have the money to grow your team.

So I'm not going to waste your time talking about how to create a pitch. Instead, the goal of this section is to give you the data points you'll need to build your pitch around. Get clarity about what engineers care about and you'll be ready to talk to them about your company in a way that gets their attention.

Never forget that you're helping your prospective hires make a hard decision. When you offer someone a job, you're essentially asking them to make a years-long commitment. It's true, if they hated it, they could leave in a few months, but most people consider short employment a serious setback and a blot on their résumé.

Good engineers have choices and will be making a tough call, and they'll want answers to key questions I'm going to share with you now. Let me quickly add that candidates don't ask these specific questions directly. These are the questions one step below their surface questions, and when you can answer them, you're getting to the heart of what candidates really care about.

My recommendation is to work through these questions and come up with answers that you internalize. That way, you'll be prepared to answer a huge range of surface questions.

- Where will they learn the most?

- What will look good on their résumé?

- Where do they have the most potential to grow?

- Where will they have the most impact?

- Will they enjoy working with your team?

- What will they be working on in the future?

- What do I tell my parents when I describe this job to them?

- What's the compensation?

You should be mindful of the answers you give any prospective software engineer. Here are some guideposts to use as you reflect on answering each of these questions.

Where will they learn the most?

Most startups are not solving "hard technical problems." Most are focused on solving business problems, so here's what you need to find out: what are the interesting technical challenges that must be overcome in order to solve the business problem?

To figure this out, sit down and talk with your small core of current engineers about the technical problems that need to be answered. What topics are they most animated talking about? What makes their eyes light up? Those are the technical problems you need to understand. Repeat it back to them until you have a firm grasp on the technical problems and can explain it to others. I recommend having three problems to be able to talk about with prospective engineers.

Here's an example from when I worked for Main Street Hub (MSH). MSH managed social networks (Facebook, Yelp, Twitter, etc.) for local businesses. They did this by having writers manually create social content and reply to comments. From a tech perspective, MSH had integrations with the social networks to automatically read and write data to those networks so that MSH writers wouldn't have to manually log into each social network for each company.

At the time, if you were doing anything with social networks, you were an attractive company to engineers, so we had that going for us. But what really attracted the engineers was the excitement of how we were going to leverage all our social data to drive new customers to our local businesses, scale the technology to handle all the customers they had to support, and learn from the social media content to build future products such as websites, automated email marketing, and customer acquisition programs.

You need to figure out similar technical challenges for your own company so you can discuss them intelligently with any engineer you meet with.

What will look good on their résumé?

You're a startup. No one knows who you are. The big players like Google, Facebook, Amazon, and Microsoft will beat you hands down for résumé recognition.

There are three pieces of good news, though. One, most accomplished and skilled engineers don't worry much about a particular job looking good on their résumé. They're secure in the knowledge that they can get the job they want based on their skills, so they're not relying on having a universally recognized name on their résumé.

Many times, younger engineers do care about getting a big name on their résumé, so the second good news is that the ones who make this a priority probably aren't the right fit for a startup.

The third piece of good news is that résumé building isn't as important as it used to be. For qualified software engineers, finding a job will be the least of their problems. It's choosing the right one. This is where you want to truly have their best interests in mind and take on the role of a career mentor.

Walk through the other companies they are thinking about and ask these same questions you're answering about their other options. You'll have zero credibility if you reflexively talk down all the other options. Instead, be a trusted partner for them and help them think through their choices. As a bonus, talking through all this will give you insight on what matters to them.

Last, if they're still concerned about their résumé, you should point out that you're building a company that will be recognized someday and here's an opportunity to be part of making that happen. And if they're still concerned about résumé building, assure them that résumé building is what our parents had to do. As a software engineer, the focus should always be on where you are going to grow and learn the fastest, because getting a job will never be an issue.

Where do they have the most potential to grow career-wise?

Established companies usually have clearly defined career paths that show recruits exactly how they can grow. You don't. But that doesn't mean you don't have advantages when answering this question.

Startups offer the potential for accelerated career growth and no predefined constraints. Any engineer you hire has to be able to handle the uncertain career path of a startup. As a tradeoff for that ambiguity, there's the chance for faster growth and harnessing their financial success to the company's rise.

Look for people who can think "company success = personal success." That's truer in a startup than a more established company because of stock options. Emphasize this theme during the hiring process: when the company is successful, their rewards go up, too. The ones who light up at that idea are the ones you want.

Where will they have the most impact?

Impact can have multiple meanings for engineers. One meaning is impact on the company. You're way ahead if that's what they care about most. If they choose an established company where they're one of a thousand or ten thousand engineers, how much impact can they have on the company? Pretty much zero; they'll be a cog in the machine at a big company. But if they come to a startup and are one of fifteen engineers, the chance for impact goes up dramatically.

Some are focused on what will have the most impact on *them*. Drill down and find out what they mean by that. This could go back to the first question about where they will learn the most. Keep going and find out what makes them tick, and then see if you can align their interests with the job you can offer them.

Will they enjoy working with your team?

This is pretty simple. They either will or won't, but you have to get them the data that helps them make that decision. We'll cover how to set up the interview process in Chapter 2, but for now, understand that you're not just interviewing them; they are also interviewing you and your team.

You need to set the process up in such a way that they understand whom they'll be working with and why they would enjoy it.

What will they be working on in the future?

This is related to the "where will they learn the most" question above, but now you want to talk about what is around the corner for your business. There may be a handful of features or new products you are dying to build. Let your excitement around these ideas be infectious.

This answer doesn't need to be extensive, but it's good to have thought about it ahead of time.

What do I tell my parents that I do?

This one is a little subtle, but it's great for thinking about how to connect with engineers. We all want to impress our parents and be able to explain to them what we do in a way that makes them understand and appreciate it.

Think of your pitch and the questions you ask them as a way to help them "tell the story" of this job to their parents. No matter how old we get, most of us still want to be able to explain what we do to our parents and have them be proud of us. (Of course, parents here are also representing all the other people in their life they love and whose respect they want. It could be a spouse, a friend, etc.)

I always ask prospective engineers at Rev, "What is it about creating work-from-home jobs that excites you?" I ask them this question in multiple ways, and one of the things this does is help them restate the mission in their own words. By having them connect and align with your mission and then giving them practice at expressing it, you make it easier for them to communicate it.

We'll revisit this more in Chapter 3, but for now, remember that you always want to help them create clarity around what the job is and why it's a fit for them. The more clarity they have around why they would work for your company, the easier they'll be able to tell the story of why they are coming to work for you.

What's the compensation?

There's a reason this is last on this list. Obviously, compensation is important, but you need to get alignment before worrying about money. If they are a fit for your mission, you can either work out the compensation or you can't. But if they aren't a match for your company, compensation is irrelevant. If they're not a match, you shouldn't want them for any amount of money, and vice versa.

The nuances of handling compensation questions will be developed more in Chapters 2 and 3.

You probably already understand that finding great engineers and getting them excited about your company is difficult. That's why it's so important to go through the steps of creating a mission and exploring your own motivations. And it's why you need to look at what engineers care about from their perspective. Finding great engineers will still be hard, but it will make you better at it.

Once you have your mission and pitch internalized, there's one more piece of the puzzle to think through: your personal motivations.

PERSONAL MOTIVATIONS

When I consult with founders, I always ask them things like, Why are you running this company? Why are *you* the one to carry out your company's mission? I encourage them to drill down, asking why again and again until they uncover the core reasons.

Examples of honest answers I've heard:

- I want to build a company that other companies are excited to see on people's résumés.

- I had a terrible boss and I never want to be in that position again. I also want to be the kind of boss that never gives my employees the same terrible experience.

- I want to be super wealthy and in control of how I make that money.

Those are all sincere answers. The last one is especially honest, although it also points to a deeper reason to uncover. Asking why wealth and control are important to you might lead to a more vulnerable answer.

Perhaps you already have a lot of clarity around your personal motivation. Still, it can be useful for your recruiting to explore it further. You'll want to be able to communicate it clearly and openly to prospective engineering hires.

Remember the goal here is to find engineers who are the right fit and give them reasons to choose your company over other offers that may also be enticing. The key is to connect with people on a personal level, because once a deep connection with another human being is made, it's a hard thing to break. They'll want to come work for you and not leave if you can create a bond. These kinds of deep connections are built on trust, and the surest way to earn someone's trust is honesty and vulnerability.

When you have a handle on your personal motivation and can communicate it, don't hold back when talking to potential hires. Make it part of a sincere conversation about who you are and why they should work for you. Obviously, forcing it or grafting it unnaturally into a chat is a bad idea. But there will come some point when you're talking to bring it up, so don't hide it. The connection you make will matter when the time comes for the engineer to make a decision.

Being transparent about who you are and why you're building a company will inevitably attract employees who are aligned with the mission you want to accomplish. Not many other people hiring software engineers are thinking this way.

Now that you have a mission, a pitch, and you've clarified your personal motivations, it's time to start finding engineers to talk to. Where do you find them?

Chapter 2

RECRUITING STEP BY STEP

When it comes to building a team, many founders immediately think "hire a recruiting agency." I know this because I've heard it directly from a lot of founders.

They tell me they've hired a recruiting agency to find and qualify candidates. Then their team will do a technical interview, and at the end, the founder swoops in to give the offer and close the candidate.

I completely understand why this sounds appealing. Founders are incredibly busy and this method optimizes for their time. However, I've seen over and over again that this method closes fewer candidates.

Even when candidates do get hired using this process, the founder hasn't taken the time to lay the foundation of a relationship with the engineer, which can lead to losing them sooner rather than later. I've also seen this method look great at first, but in six months, most of the new hires are gone because the founder didn't have time to properly evaluate the candidates for the proper fit.

I'm not saying that recruiting agencies are always a bad idea, but if you see them as a shortcut, you're probably going to be disappointed. There's a better way to do that, and the best way to see it is to start by understanding your hiring funnel.

The simplest way to conceptualize adding engineers to your teams is this: it's exactly like a sales funnel, except the bottom of the funnel is not a sale—it's the right hire. When you see recruitment through this lens, the things you need to accomplish become clearer.

For example, the idea of a sales funnel is to pour as many qualified leads as you can into the top of it. Same thing with recruiting.

You want to cast as wide a net as possible to fill up the top of your hiring funnel with qualified engineering prospects.

You'll of course want to use your network to add candidates to the funnel. However, it's good to remember that others are doing this, too, so there are limits to how well this will work. In my experience, you'll get one or two good hires from referrals in your network but not nearly enough to build a team.

The first thing you need to do is establish a baseline for who qualifies for the top of your funnel. You can (and should) keep the criteria somewhat flexible, but you do need something more specific than "all software engineers." The way to establish the criteria is to go to your core of current engineers and say something like, "If you could wave a magic wand and have three engineers start tomorrow, what would be their ideal work skills and background?"

They'll give you some criteria like years of experience, the technologies the person has worked with, and so forth. Use that as the basis when searching, but take the perfect criteria they give you and make them looser.

Example: If the ideal is five years' experience, make your actual search criterion two to ten years' experience. Do that with everything they give you. Remember, this is about a wide net, pushing as many potential hires into the top of the funnel, while still staying within shouting distance of the criteria.

Once you have defined your wide net, where do you cast it?

Number one is LinkedIn. I recommend spending an hour a day sending messages to potential hires. Keep your message short, simple, and to the point: introduce yourself and your company, describe the position you have available, pique their interest on what your company is working on, and ask them if they'd like to meet.

Here's an example:

> Hi, I am the founder of the startup Code4All. We are creating a new platform that allows anyone to build sophisticated mobile apps using your mobile device.
>
> We are looking for a few engineers who are interested in building a new company from scratch and helping us figure out how to put the power of creating apps into the hands of an average person.
>
> I would love to talk to you about the different opportunities here at Code4All. Would you be interested in learning more?
>
> David, CEO/Founder

This is just an example; I recommend testing different messages to find out what generates the most responses. In general, make it as frictionless as possible for a candidate to read and respond. Once the conversation is started, there'll be time for more details.

If you spend at least an hour a day, the responses will build. They'll accrue because many won't respond immediately. By doing it every day, the cumulative effect will start filling up the top of your funnel. To give you some rough numbers from my experience, about 20 percent of people will respond. Of those, half will want to meet, and then half of those you'll actually end up meeting. In other words, about 5 percent of your LinkedIn outreach will turn into a meeting. If your numbers are significantly below these benchmarks, take a hard look at the message you're sending and test some adjustments to see if you raise your response rate.

This LinkedIn strategy is the best way to get engineers into the top of your funnel, but here are few more to consider.

Post your job on websites like Indeed. This shouldn't be your go-to strategy for recruitment, but because posting on big job sites doesn't take much time, it's worth doing. Besides getting some additional people into your funnel, it puts your company name out into the world. If an engineer sees it randomly or searches for the posting after you've reached out on LinkedIn, it's one more way to plant the seed of working for your company.

This is also handy because it can give you something to refer people to if they ask for a job description. Be sure to post the job on your own website, too.

Talk, talk, talk. Talk about hiring engineers. You likely have connections with other founders and tech investors, plus a network of friends and acquaintances. Whenever it fits naturally into the conversation, make a conscious effort to talk about what you're looking for. They may have ideas on recruiting, or even better, people they can directly refer to you.

College recruiting. I mention this because you may be wondering about it, but my advice is to skip this one for now. It can be a great source of engineers later in your company's growth, but it is a long, arduous process. It typically takes at least a year to land them. If you're looking to get your next five to twenty engineers hired relatively quickly, college recruiting is not the way to go. (However, if you get to a point where you're ready to recruit from colleges, put your team on it sooner rather than later. It takes years to be successful at colleges, so start the clock on that when you can.)

Once you begin getting people into the top of the funnel, how do you qualify them to see if they should move on down the funnel? You as the founder need to meet with them yourself, preferably face-to-face.

BEER OR COFFEE?

This first meeting should be in a casual setting, so I recommend inviting them out for a coffee or a beer. What you're doing in this first meeting is trying to build a relationship, and a formal office interview is not the right setting to start a relationship.

Also, don't refer to this as an interview. Call it a meeting and describe the purpose to the engineer along these lines:

"I'd like to meet with you, just to answer any questions you have, tell you a little about us, and see if we have a potential fit on both sides."

Simple, casual, and pressure-free. A relaxed atmosphere is your best chance to connect, and if you don't, you likely won't be moving to the next step with the candidate.

I recommend you give yourself four tasks at this first meeting:

▸ Clearly state your mission and pay close attention to their reaction.

▸ Share who you are and why you're doing this.

▸ Test and make mental notes on how to refine your pitch.

- ▶ Determine whether this is a person who should move to the next step in the process.

The first three points were detailed in Chapter 1. It's now, at these casual one-on-one meetings, that your preparation pays off. All the thought you put into creating clarity on mission, personal motivation, and your pitch will be used in these meetings. You're ready with honest, well-thought-out, and vulnerable answers that can create deep connections.

You also want to use this as a chance to make mental notes to improve future pitches with candidates. For example, when you're talking about things like "where will you learn the most," what answers connect with candidates and which are misfires? It might be only this candidate who is not sparking with it, but it also might mean your pitch needs adjusting.

Last, in these conversations, make sure you ask questions and then get out of your own way. Make sure they are talking more than you. Don't interrupt with the positive points about your company when they say something that triggers a connection to your pitch; just make mental notes. If they need clarification or ask questions, of course, answer. But once you listen and understand what drives your candidate, thank them for being so open and then it's your turn. This is your chance to match their needs with what you have to offer.

SHOULD THIS PERSON MOVE TO THE NEXT STEP IN YOUR INTERVIEW PROCESS?

From a practical standpoint, the critical answer that needs to come out of this meeting is this: is this a person who should move forward to the next step in your hiring process?

This of course is not an exact science, and many times, it will boil down to your gut reaction of the entire experience with the person. Nonetheless, there are some general guidelines to follow.

One of the first questions I ask myself when deciding is, "Are they nice?" I'm sure some will read that and say it's a stupid thing to judge an engineer on. After all, being nice isn't a requirement for knowing how to write code. Strictly speaking, that's true, but you're looking for people who can bond with your mission and who care about others on the team. In my experience, anyone who is arrogant or abrasive won't work in that kind of company culture.

Here's something else I ask myself when deciding if they should move to the next step. "Do they communicate well?" I find this out by asking them questions like:

▸ Why are you considering changing jobs (and/or why did you change jobs in the past)?

- ▸ What does your current company do?

- ▸ Tell me about the kind of things you've worked on in the past for other companies (or what are you working on at your current job)?

When they explain job changes, the most basic thing I'm looking for is honesty. The answer I most like to hear is "I can't grow there anymore." If their current company is stagnant and their manager isn't likely to leave anytime soon, how can they grow? This is tailor made for startups, where the potential for growth is phenomenal.

Another answer I hear a lot is "There's no more interesting work." This is usually a good answer. If their current job doesn't offer them interesting work, this is where you can see if they get excited about the technical challenges your company needs to solve. But this can also be a potential red flag answer. Is this the type of engineer who gets bored super easy and super fast?

Explore this by asking more questions and looking at their résumé. If they've had very frequent job changes without clear reasons, that's a sign they get bored easily. If they grow tired of any kind of coding work and need constant freshness, that's a hire who will leave you too soon.

When I ask about what their current or past companies do, I simply want to know that they know it and can communicate it reasonably well. You wouldn't believe the number of engineers who don't know what their company does! If they don't know it or can't communicate it, that's not someone I want. I want people who care.

Understand that when you ask about what their current company does, you'll sometimes get very technical answers. Let the person finish, but then ask again. It's rare, but the best engineers will directly answer with the mission of the company. Others will answer with what their company sells. This is at least better than not knowing at all. What I find is that great engineers know what they are doing for their users because they care about the users.

When I ask them to explain the kinds of things they've worked on at other companies, my main interest is whether they can explain it in normal human terms. I'd also like to hear them say why what they did was important to the company. Again, I want people who can communicate well, not just write code.

Another thing to listen for is too much negativity. There's a balance here. Sometimes the candidate is expressing legitimate concerns and criticisms with their current job. But there are times when you find someone who you sense is going to be unhappy and negative anywhere they are.

I remember one time at Rev having the perfect candidate on paper to head up a new division for us. He had experience working at Amazon and his background was perfect. But his negativity during the interview was so strong that I had to decide against hiring him. A tough call, but I think the right one.

Here are some more things I recommend probing for answers on so you know if they should come in for interviews:

- ▸ Do they care about their craft? If they do, it will come out naturally as you talk. Listen for it, and if it's absent, they won't be a good hire.

- ▸ What other companies are they talking to? When you ask this, help them by walking them through how it compares to what your job offers. Listen closely for what they care about. Then show them you care about them by giving them honest pros and cons to their different potential options. Take on the role of a mentor helping them by truly having their best interests at heart.

- ▸ Bring up your company mission a lot and gauge the response. Do they seem to be regurgitating what they think you want to hear, or are they genuinely excited about it? Ask open-ended questions about the mission and see what they say. Remember that part of what you want is for them to be able to repeat to parents, spouses, and friends what their new job is all about.

If you decide you are interested in moving them further down the funnel, the final thing you need to ask about is compensation. (If you're not interested in hiring them, there's no reason to bring it up.) You are not looking to go into this in depth at this first meeting. Ask, "What are you targeting from a cash perspective?" If it's in line with the ranges you have in mind, leave it at that.

Keep in mind that you'll usually have to push people on this. Candidates will often respond that they don't know or are just looking for the best you can do. This is where you need to be clear on how cash is precious. If you still can't get a number, you can mention that you don't want to waste anyone's time, and if their cash needs are out of scope for you, it might not make sense to move forward with the interview.

I can tell you I've failed at times to get a salary range with specific candidates. It almost always comes back to bite me later, so I strongly recommend working hard to get this information.

Once you do get a number, if it's too great a mismatch on cash compensation, it's probably a waste of time to take it any further.

Some people have big mortgages to pay, private school bills for kids, and so on, and no amount of stock options can help them.

However, there will be many cases where either they won't be specific about a number, or they give you a number that's within shouting distance of your range. Don't start negotiating at this first meeting. But you can talk in general terms, something along these lines:

"I'm going to be honest. At startups, cash is precious. If you get an offer from me and an offer from Facebook, mine will be lower. What we can give, besides a great job, is ownership in the form of stock options. It's the difference between owning a house and renting a house. When you own, you really care about it."

From there, if they haven't given you much to go on, you have a few options. First, you can say, "Take some time and let's jump on a call in a few days to revisit before moving forward with interviews." Second, say something like, "If I was able to get in the range of X [whatever you can pay] with stock options, would you be interested?" Then sit back and listen. I recommend against suggesting an amount yourself because that tends to lock you into a number.

The goal on compensation in this first meeting is simple: see if you can get a range from them, and then figure out if you're close enough that it could work.

If the prospective engineer is excited about the mission, is nice and a good communicator, and the compensation needs are in line, it's time for the next step.

THE INTERVIEW PROCESS

Your formal interview process will be dependent on your specific company, but I consider three interviews to be ideal.

First, have at least two "technical" interviews that last about ninety minutes. These are to determine if they can do the actual coding work they'll need to do the job. Then have one sixty-minute interview with the manager they'll be reporting to. This interview is not technical and is aimed at ensuring the fit between the prospective engineer and their manager.

Some companies will want to do more interviews. That's fine, but I wouldn't let it be more than five interviews. At a certain point, an excessive number of interviews just makes your process look weak and disorganized.

Design the technical interviews using input from your current engineers. What questions and tests can you give candidates to determine if they have the necessary skills?

Be sure to remind your interviewers that they are both qualifying the candidate and selling the company. If this is a great candidate, you want your team to demonstrate this will be a great place to work and a chance to master their craft.

The only way to truly know if your interviewing process is good is to do it and find out. If at the end of the interviews, your team is saying, "X candidate did really well, but I don't know if I want to hire him," your process is flawed. You need to figure out how to fix it. It could be that the technical questions weren't hard enough so you don't know what they can do. Also, your questions may be too specific. You want questions with a little ambiguity in them, which will force the engineer to ask clarification questions in response.

There's one other important thing you should make sure your interviewers understand, and that's what to do if it becomes clear a person cannot do the work. Always be kind. Don't make them feel like they are wasting your time. I've had some of my best referrals come from people I've said no to.

Your goal for the interviews should be to create consistency in your process. The interviews should be distinct, but the first interview should be the same for every candidate, and the same with the second interview. You can evolve your questions over time, but in general, you should standardize your questions for your interviews.

The manager interview should be focused on figuring out if the person would work with the team successfully. A solid approach is to walk through the résumé and ask about challenges the candidate has had in each work situation. A manager should listen for answers that demonstrate the engineer can communicate on both a technical and nontechnical level.

In the best scenario, they do wonderfully during the interviews and your team wants them on board. It's time for you as the founder to take the process back over.

Chapter 3

CLOSING ENGINEERS

You're now at the key moment at the bottom of your hiring funnel. Much of your company's success will be determined by getting the right engineers to say yes.

The good news is, if you implement the concepts in Chapters 1 and 2, you're likely about 80–90 percent of the way to closing and getting them hired. Here, again, it's a lot like sales. When you're selling something, you're checking all along for alignment and understanding with the prospective buyer. When done correctly, the close becomes the next natural step in the sales process.

In the same way, the first two chapters were about laying the groundwork by aligning your prospective hires with your mission and its technical challenges. If you've also created a personal connection, the closing—getting that yes from the engineer—will happen much more easily.

Let's dive into the practicalities of closing. First, it should be personally handled by the founder. As you grow and add people, that won't always be the case, but for these first key hires, it should be you. The setting should once again be a casual atmosphere over coffee or beer.

The best attitude to bring to this final meeting is as a mentor. You should be helping them think through a key life decision and truly have their best interests at heart. Let's walk through an ideal scenario of how this meeting should go, and then we'll talk about situations where everything doesn't go perfectly (which of course happens all the time).

After exchanging some pleasant small talk, communicate to them how well they did in the interview process. Something along the lines of, "My team loved you." Add a few personalized, specific details as to why.

Next, talk about impact and growth in their craft. "We think you can have a massive impact here. You can grow here by helping us overcome challenging technical problems and at the same time continue mastering your own craft." Obviously, you want any compliment you deliver to be sincere and supported in fact. If you don't fully believe what you're saying, you shouldn't be offering them a job.

Next, you want to hear from them. Ask, "What do you think about the company?" and then *listen*.

In many cases, what you'll hear back is obvious excitement. That's because of the foundation you and your team put in place prior to this meeting. In other cases, the response will be anywhere from strong interest all the way down to tepid. This is when you need to listen carefully to what they say. If it's clear they aren't going to get excited about your mission, your company, or your team, it might not be the right fit.

But in some cases, the candidate is truly conflicted about whether they should work for your company. In this case, find out what their concerns are and see if you can remind them of the alignment between what they want and what your job offers.

Now it's time to ask them the key question: "If we can figure out a compensation package that works for you, would you come work here?"

In the ideal situation, you'll get an immediate yes. This is gold because it's going to be very hard to give a final no later. They want to work for you, and now it's a matter of getting the compensation to match.

If they waffle on the compensation answer, it's time to dig down and find out why. In a few cases, it's that they have already decided against the job but are the kind of person who has trouble saying no.

Some just want to wait to hear the actual compensation offer before saying the word "yes." Other times, you won't be able to uncover the exact reason.

If you get a yes, or at least not a no, it's time to deliver the compensation offer verbally. "This is what I'd like to offer you." Give them the dollar amount and stock options. Then observe closely.

If you see or hear excitement, you're good to go. However, if you sense disappointment, don't ignore that. Bring it to the surface. "I notice that you are a little disappointed. Talk to me about why."

One common reason for ambiguity is they want to weigh other offers, either ones already on the table or potential. When you get this type of answer, you have to handle it on a case-by-case basis. If you're starting to sense that maybe you misread them and they weren't as excited about your mission, the opportunity, or the company as you first thought, maybe they aren't the right fit after all.

Other times, they truly aren't sure what's best for them and their career. In these cases, you should be a mentor to them and help them talk through their other potential offers. What are the pros and cons of each? If you are relentlessly negative about other possibilities, you're doing two things wrong. One, you're not truly caring about the person in front of you. Two, the engineer is likely to pick up on your lack of objectivity. It can undermine everything you've done previously to create a genuine connection.

Ultimately, if they are weighing your offer against others, your best answer to them is, "I want you to get your other offers so that you can have a full picture of your opportunities. Let's get back together once you have all the data." You've done what you can, and if the person doesn't end up working for you, it wasn't meant to be. If they had a burning passion for your mission, they'd already be on board.

There's another reason that is common and obvious: the engineer actually wanted/expected more money. In the perfect scenario, this wouldn't happen because you'd be offering them something within or close to the range they had specified in their first meeting with you. But miscommunications, changing circumstances, and other situations happen. Whatever the reason, the money isn't in alignment.

If it is just a case of mismatched compensation, find out what they are looking for by asking them directly. If the number they cite is not out of reach, I recommend some version of this: "It's not what I was prepared to offer today. I need to circle back with my accounting people (or CFO) and talk it over with them. If I could get it to 'X' dollars, is that a yes?" If it is, give it twenty-four hours and reach out with an updated offer. Whether it's for the full amount they want or some kind of split-the-difference situation is something you'll need to decide.

Of course, stock options are part of the compensation equation. Chapter 2 covered how to present the balance between cash and stock options (cash is precious and stock is ownership), so we won't go over it again here. But you should keep it in mind and be comfortable reminding them of the upside of stock options, while being honest that it's not a guarantee. If there is a compensation difference, explore if increasing stock options could make up for a gap in the cash they would like.

In the end, when it comes to compensation, the ideal engineer will understand that cash is more precious at a new company. They'll be used to options being part of the package and appreciate the added satisfaction of having a huge impact on a company. If the person can't get past a lower compensation for a potentially larger upside and more job satisfaction, they're not a fit for a startup.

Let's briefly cover two nonstandard situations. Sometimes you'll come across an engineer who is excellent but still is overpaid compared to industry standards. There are many reasons why this could have happened, but the important thing to grasp is that many engineers care about the work and company goals over cash. In other words, don't necessarily rule this person out.

Second, there are of course times early on when you may want to overrule your own compensation guidelines to get a person you really want. As a founder and entrepreneur, you know there are always exceptions. Make it rare, but there can be times when you'll want to bend your own rules.

Despite everything you do, there will be times when someone you want doesn't close. Never be bitter. Sincerely congratulate them. Here's what I say: "Congratulations. I am so happy that you found something you are excited about. Let's please stay in touch in case something doesn't work out or if you just wake up one day and say, 'I really want to work with you guys.' There are no bridges burned here. Also, if you know someone who you think would be interested, please connect us." Always be looking for ways to fill that funnel.

Of course, other times it will work out great. Then it's time to . . .

CELEBRATE

When you get a commitment, spring into action. First, immediately talk about the paperwork you'll be sending over for them to sign, and then email it that same day or evening. Every engineer feels more secure receiving that paperwork to seal the deal. Get it to them fast.

Also, as soon as you get that yes, ping everyone in your company who was involved in the recruiting and interviewing process. They should all reach out with congratulatory messages, which will reinforce to the hire that it's a great team.

You should also get creative and figure out ways to add to the welcome. It can be as simple as overnighting a company T-shirt to them or maybe sending them kid swag if they have children. The exact recognition you choose to do is not as important as that you do something. For almost everyone, it's never just money, and these little touches matter.

This is a moment that deserves to be celebrated all around, of course. You've found a great engineer for your company, which as you can see, takes some work. The last thing you want to do is lose these great hires you're recruiting. How do you keep them on board?

Chapter 4

KNOW YOUR ENGINEERS

What does losing a good engineer from your team cost you if you're still in the early stages of your startup?

According to the Society for Human Resource Management (SHRM), the average replacement cost for salaried employees is six to nine months' salary. For a person making $100,000 a year, you can expect at least $50,000 to $75,000 in replacement costs. (It should also be noted that there are alternative studies that say the number is much higher, sometimes as much as double the annual salary in replacement costs.)

However you quantify it in money terms, I believe that no metric can completely capture the true cost. For one thing, you've seen over the last three chapters how much goes into finding and hiring engineers who fit your company and mission. It's demoralizing to see all that effort walk out the door.

But even worse is the amount of the company's intellectual capital that vanishes the moment they leave. Let's say this is an engineer who is part of the first wave of key hires we're talking about in this book. There's no one who you can hire that will have the depth of specific company knowledge as the engineer who's walking out the door. Even if you find an engineer with the equivalent experience in industry terms, it won't be experience at *your* company.

It's crucial to find ways to keep your engineers excited about the work and connected to you and your company. I recommend continuing the relationship you started building during the hiring process. Go deeper in understanding what makes them tick as individuals.

There are two essential parts to this concept of knowing your engineers.

Step one is making sure you understand that knowing your engineers is not a surface thing. Here's an all-too-typical example of an interaction between founders and their engineers:

Founder: "How are you doing?"

Engineer: "Oh, pretty good."

Then maybe a few more pleasantries are exchanged, and it all goes nowhere.

If that doesn't work, what's a better way? Start by ridding yourself of a caricature of software engineers that is all too common. The usual idea is something along these lines:

"Engineers are introverts. They live in a world that I can't understand (the mysterious world of writing code). There's not much to be gained from trying to connect with them or understand them—their world is too insular."

It's human nature to think like this. There are many other professions where people make assumptions based on the mysteriousness of the work being done. (Mathematicians, genetic scientists, and surgeons come to mind, and you can probably think of a few examples of your own.)

This kind of thinking leads most people to put software engineers all in one basket. That basket is labeled "I don't understand what software engineers do, and therefore they won't be relatable to me." That's exactly wrong.

I know firsthand it's not true because I've hired and managed hundreds of engineers over my career. Lumping together all software engineers as mysterious introverts with awkward social skills is false. Of course, just like with any group of people, there are some who are introverted and harder to get to know. But that doesn't mean you can't connect with them, too.

If you convince yourself that "I'm not going to be able to get on their wavelength," it will turn into a self-fulfilling prophecy. You'll get back what you give, and they'll all remain mysterious to you.

But if you make the effort to get to know your software engineers as individuals, your view of them will change, and so will your idea of what will motivate them. What's the best way to get to know them?

At its most basic, it's about setting aside the time. As the founder, your personal attention matters to them. When the company leader takes real interest in a person, it makes them feel valued.

What are the best questions for engagement? As I mentioned already, the overused greeting "How are you doing?" is going to end up nowhere as a conversation starter. Here are three questions I've found to be more effective in creating a true dialogue:

- "What's on your mind?"

- "Where are we failing you?" and/or "How could we be supporting you more?"

- "Are you working on things that are making you better?"

Whether you use these specific questions or not, I'm assuming as the founder of a company who has launched and raised money, you know more than a little something about connecting with people. Bring those same skills when you engage with your engineers.

There are several other key points about getting to know your engineers. First, do not spend all your time talking about work-related issues. In fact, spend more time talking about other subjects. Find out what your individual engineers do in their spare time and what they enjoy talking about. The setting for these conversations shouldn't always be work. Take them out for coffee, or a beer, or a lunch or dinner.

Don't let yourself be distracted when you're with them. Stop checking your phone or otherwise allowing your attention to be stolen. People notice this and it sends the message that you aren't really interested in them.

It is of course appropriate to spend some of the time (not all!) talking to them about work. When you do, make a mental note of what they want to be working on in the future. Imagine how it makes them feel if later you remember that and make that opportunity available to them. Actions like that bond them to you and the company.

Last, make a real effort to understand their goals, and not just work-related goals. You will find that many engineers just want to be better engineers. If you ask a doctor what they want to be in five years, the answer is often "a better doctor." It can be the same with engineers. What engineers mean is that they most want to be able to solve harder problems with autonomy and with greater efficiency.

Outside of work, they have the same kind of goals as most people, with hopes for their families, sports, travel, or having their own businesses. One time, I had an engineer who told me that he was saving money to eventually take time off and travel the world. I was able to offer him a way to do that while still doing high-quality, productive work for us and not having to quit his job. Without taking the time to learn about his personal goals, I never would have known to offer him that chance.

It's definitely worth investing the time to listen on a personal level. You will find ways to help align the flexibility of their work to help them meet personal goals.

THE FRAMEWORK OF
PURPOSE, AUTONOMY, AND MASTERY

People everywhere want to find fulfillment and meaning in their work, engineers included. That, more than money, keeps people from leaving a job. But what exactly drives fulfillment and meaning? Daniel Pink explains it brilliantly in his book *Drive*. People look for purpose, mastery, and autonomy. When you give your employees the opportunities for all three, they love what they do, and they have no desire to leave.

Here are some guideposts for implementing each of these concepts for your engineering team.

Purpose

You've got an excellent head start for creating a purposeful work-place because your company has a mission. And if you followed the advice in the recruitment section, you hired engineers who are energized by the mission. You need to extend and reinforce the momentum of the mission throughout the day-to-day work-place events.

Your mission should take place front and center in company meetings, initiatives, and informal conversations. It's particu-larly important that you as the founder reinforce it by talking about it a lot and keeping your team pointed at this company North Star.

You might come to a point where you feel like you are always repeating it and be tempted to stop. On the contrary, that's a sign you're doing it right. While you are hearing yourself say it each time, the people you are saying it to are not. To them, it's not overly repetitive because they aren't there every time you say it.

Make sure when managers assign work, they aren't just saying, "Do this." They should be putting tasks in the context of the mis-sion. "We need you to do X because it will allow us to [fulfill this part of our mission]." Keeping mission front and center gives people the purpose they crave.

Look for ways to align engineers with particular parts of your mission. The key is to match up people with the aspect of the mission they most care about. How do you know what fires up an individual engineer? That's where the conversations you have with them come in. You also need to encourage your managers to have these same kinds of conversations regularly and share notes about which engineers resonate with which part of the mission.

Here's an example of how this plays out at Rev. We have teams working on all parts of our platform. As a marketplace, we have two main users, customers and freelancers. The more we invest in getting customers, the more work-from-home jobs we can create. The more we invest in our freelancers, they perform better work and in turn we have happier customers. Some engineers really enjoy working on customer problems, and others love working on freelancer problems. Sometimes you have to make a tradeoff on how much you invest in parts of your business based on the desires of your team. There is a healthy balance here. I can't have all my engineers working on customer problems. If you are transparent about what your company needs to accomplish and show your team that when opportunities arise where you can align people with their passion, you can create a culture where engineers are happy to work on almost anything.

What you'll find is that fulfilling the mission has several aspects, and these distinct parts resonate with individual engineers differently.

Mastery

Legendary college football coach Lou Holtz had a reputation as a master motivator and teacher. But when he was asked one time about how he motivated his players, his answer was interesting.

"It's not my job to motivate players," he said. "They bring extraordinary motivation to our program. It's my job not to demotivate them."

One of the best ways to kill motivation in a driven engineer is to keep piling mundane tasks on them that never challenge them. And you can finish off the job of crushing their motivation by never explaining why they're working on something.

Of course, it needs to be said that there are mundane tasks that need doing. Colors need to be changed on a website, images need resizing, buttons need redoing, and text must be altered. That can't be ignored.

How do you balance the need to get mundane tasks done with the necessity of giving your engineers the chance to continue mastering their craft?

Part of the answer is to add a challenge into the run-of-the-mill tasks. I tell my engineers that practice makes perfect if you do it the right way. Can they challenge themselves to find more efficient ways to complete these otherwise boring tasks? They can get better at even these tasks if they push themselves to do it. Like a basketball player shooting free throw after free throw, there's a feeling of accomplishment when you get better at it.

Even more importantly, you need to give them a mix of the routine along with coding projects that stretch their skills. You don't want to overdo it and give someone something they're sure to fail at, but you do need to allow them to push past their current limits.

The crucial concept is for you and your engineering leadership team to be very conscious and intentional about the mix of tasks you give each engineer. At many startups, this idea gets drowned by having too much to do. Everything ends up being "we need this done now." Yes, things need to be done, but if you keep giving the same engineers nothing but a list of boring coding tasks, many will leave at the next good offer they get.

Give them opportunities to continually improve and master their craft.

Autonomy

People are happier and more satisfied at work when they are given autonomy. There's always the temptation to tell employees not just what needs done but also exactly how to do it.

Engineers do need to work within certain structures, but as much as possible, you need to work closely with them to define the problems and the solutions.

To implement as much autonomy as possible, you'll need to have clear definitions of what "done" means for any particular task, and you'll also need a culture of healthy feedback. We'll cover that more in the next chapter.

There will be times when complicated architecture will require more oversight than typical. But even in these cases, don't have someone looking over their shoulder constantly or telling them exactly how to do it. Instead, build in more check-in points. Usually, you'd wait until the definition of done has been met before asking them to share their work. But on these more complicated tasks, have them deliver multiple work-in-progress check-ins. They keep their autonomy without things ending in chaos.

This concept of autonomy should go deeper than allowing your engineers to solve an algorithm or architecture in their own way.

It should also include involving them in solving your business objectives from the beginning. Come to them with the problem, not the solution. Engineers have great insight to how your code works and will often have great ways to solve a problem faster if they understand your long-term goals.

I recommend the book *Ask Your Developer* by Jeff Lawson on this topic. It goes deep on how Twilio's entire business was built on working with your developers to solve a problem, not just asking them to build an already-designed solution.

I consider the concepts of purpose, mastery, and autonomy to be indispensable. You should put everything related to your engineers through these three filters. I literally bring it up in almost every work conversation I have with engineers in one form or another. Give your employees purpose, autonomy, and the chance to master their craft, and they'll never want to leave.

DO YOU SEE YOUR ENGINEERS AS FACTORY WORKERS OR ARTISTS?

At the end of the day, how do you see your engineers? Are they factory workers standing on an assembly line that cranks out line after line of code? Or are they artists, using their craftsmanship to create amazing tech products?

The honest answer is some of both. Your team will create a production line that produces code. Specs are created, code is written, and solutions are built. There are many ways to optimize this process just like any production line. But if you want to retain talent, thinking of them exclusively as factory workers will lead to rapid churn.

I hear this attitude from some founders when I talk to them: "I need to hire ten engineers so I can get this list of fifteen things done." It's understandable, and those fifteen things do probably need to be done. However, the mindset of "I just need these things done" can lead to seeing your engineers as only cogs in your machine instead of the artists who design the cogs.

You need to respect your engineers as artists but also look for ways to help them optimize ways to produce code efficiently. The truth is, there's nearly an infinite number of ways to build a product. Being clear about what you need to accomplish but being more open-minded about how to get there will inspire your engineers.

Artists work with purpose, autonomy, and strive for mastery. Treat your engineers like artists, but don't completely neglect the efficiency of the factory worker side of the equation.

Chapter 5

CREATING A CULTURE OF CONSTANT IMPROVEMENT

S everal years back, I had an engineer who had been working at Rev for a week come to me with a worried look on his face.

"I've never had this much feedback anywhere in my life," he said. "Am I going to get fired?" We laughed. But behind his joke, there was a hint of concern.

This guy is a great software engineer with ten years' experience in the industry. He was in absolutely no danger of losing his job, which I assured him. I explained that's our culture. The feedback is always respectful, but it will be honest, and there will be a lot of it.

It's been six years since we had that conversation. He embraced the feedback culture at Rev and is now a leader in the organization.

That's the kind of culture you want to build, and this chapter will show you the way to get there.

But first, let me tell you what the opposite looks like.

At another company I worked for previously, there was an engineer who absolutely hated his job. He was sure the company was going to fail, that he wouldn't strike it rich, and that his job would eventually disappear along with the company. He was super bitter about it, and I found out just how bitter after he left.

Before he left, he continued to write at least some code, even if his main purpose was to buy time until he could jump ship. Eventually, he got another job and moved on. It was a relief to no longer have to deal with his negativity, but it turned out we weren't quite rid of him. He'd left a little present buried in his code.

I found it because I had to manage his code after he left the company. To understand what he did, you first need to know the conventions for naming variables in code. Engineers are supposed to name code variables in a way that gives fellow engineers an insight into the purpose of the variable.

Here's what I found when reviewing some code this engineer had created. Every variable was named "fuck1, fuck2, fuck3," and hopefully you see the pattern. It was a literal "F you" to the company. Now, on one level, it can make for a humorous war story of the "you won't believe what this engineer did" variety. I get it, and I can even laugh about it a little myself.

But it's kind of sad that he had that much bitterness. The fact that he did it in the code makes it much worse. If he had left a similar message on a whiteboard in the office or even had scrawled it on a cubicle wall, it wouldn't have spoken well of who he was, but that would've been the end of it.

But when someone does this in the code, it's not just an ugly goodbye; it's also destructive. The fact that the message was profanity added an extra bit of nastiness, but the truth is, he could've named the variables "happy1, happy2, happy 3" and it would have been equally destructive.

If you give the wrong engineer the power to wreak havoc on your code, you're giving him the keys to hurt your company and your product. As a founder, you need to understand how things can take a bad turn and why it's crucial to have the right environment.

It starts with the fact that the work software engineers do is complicated and hard for most people to understand. This makes it easier for them to hide, take shortcuts, and be deceptive than it would be in most jobs. And if you have the wrong culture, it's guaranteed that many of your engineers will do all these things.

It reminds me of when you take your car to a mechanic. You're pretty much at the mercy of whatever you're told. From the diagnosis, to the repair, to the pricing, it can be near impossible to know if you're getting accurate information.

In a similar way, you're at the mercy of your software engineers, too (unless you get the right people and build the right culture). Here are two common examples of how engineers can create problems without you knowing it:

- ▸ Exaggerating how long something should take them.

- ▸ Creating code that is just barely good enough to get a feature to work but not something that anybody else can build on later because it was done so sloppily in the first place.

I've taken some time to explain what can go wrong so you can grasp what's at stake here. Poor culture and the wrong engineers are like termites that can slowly eat away at your company without you knowing it until it's too late. Eventually, the whole thing collapses, and you'll be left stunned and wondering what happened.

The good news is, there are ways to build a culture that can make sure these things don't happen in your company and ways to inspire the opposite behavior. Instead of slowing things down or doing the least they can do, your engineers will be motivated to continuously improve.

The rest of this chapter is dedicated to explaining how to do that, but I do need to throw in a caveat here. There's not one single culture that will work for every software engineering company. I can't lay out an exact set of checkboxes and say, "If you do this, you'll have the perfect culture."

It doesn't work like that because cultures need to be built organically. It's at least as much art as science. I can give you a few common building blocks you can use to grow a healthy culture. And I can provide you some guideposts to keep in mind as early warning signs that your culture could be deteriorating.

START WITH THE RIGHT PEOPLE

The first building block is having the right software engineers—people who align with your mission. That's why this book spends the first three chapters on recruiting and hiring, and why there was so much emphasis on the importance of getting engineers who are excited about what your company does.

If you carefully screen engineers to make sure they align with your mission, do you think they are likely to be secretly lacing your code with F-bombs? The engineers you spend time talking with outside of work and showed you care about—are they the ones doing the bare minimum needed to get a feature working? Not likely.

The second building block is also tied to your mission. You want a culture that constantly ties tasks to specific parts of the mission. Let me give you a couple of examples from Rev. Remember our mission is to create work-from-home jobs.

> ▸ As I mentioned in the previous chapter,
> I have teams of engineers focusing on customer
> experience. The better the customer experience,
> the more transcripts we sell. The more transcripts
> we sell, the more work-from-home jobs
> we create.

- I have a team that figures out internally how to match the right jobs with the right people. When this team succeeds, happiness and productivity go up, creating a better product. That leads to happier customers who buy more, and that creates more jobs.

Healthy startup cultures think and talk like this on a daily basis. Tie the work back to how it furthers the mission. If you're having trouble tying tasks to the company mission, you're doing it wrong. Either you have the wrong mission or you're doing the wrong things.

The third building block to a healthy culture is building a culture of honest and respectful feedback.

The baseline foundation of any good feedback culture is an iron-clad rule to always give it respectfully. It's all about how it's delivered; that's why I emphasize hiring people who are nice. Nothing makes feedback more unpalatable than someone who delivers it like a jerk. Feedback that humiliates a person or makes them feel threatened or attacked will kill your culture faster than anything else.

However, that doesn't mean the feedback should be an easy pat on the back. That's just as bad because it gives the illusion of feedback while improving no one. Feedback must be tough-minded, but how do you get people to accept it without getting into a defensive crouch?

You explain that feedback is not negative criticism. It's actually a positive invitation to master the engineer's craft.

The only way to advance rapidly in writing code is as part of a team that emphasizes constant learning and feedback. An engineer can't go off on his own and read some manual and expect to achieve mastery of his craft. Mastery is the result of learning in an environment where there is challenging work and tough-minded, honest feedback.

Provide this environment and emphasize its importance to your engineers. They should crave honest feedback because it's the path to mastery. Instead of shying away from it or worrying that it means something's wrong with them, they should receive it with gratitude. This kind of culture works. We've built it at Rev, and I've seen it succeed in other places.

Even better, as a startup, you can choose to create a feedback culture right from the beginning instead of trying to graft it on later.

CODE REVIEWS AS PART OF
YOUR CULTURE

You may already be familiar with pull requests (PRs). In essence, when engineers submit a PR, they're requesting their code be reviewed so they can receive feedback on it prior to it going into production. Engineers giving feedback on PRs should be baked into your culture as the expectation, not the exception.

I would estimate that 80 percent of tech companies that use PRs as part of their release process don't actually have a culture where engineers are reviewing the code. You need to be in that 20 percent of tech companies that take PRs as a serious feedback process that raises everyone's game.

When this kind of real feedback is normalized as part of the culture, your engineers become used to it. They barely notice how thick their skin has gotten because they're too busy getting better and better at their craft. The bottom line on PRs and code reviews is this: immediately make them part of your culture and insist that they be more than just a rubber stamp.

EXPECT MORE THAN
JUST CODING EXCELLENCE

So far, the emphasis has been on coding feedback. I'm a firm believer that you want well-rounded engineers who work together to make the team stronger. The only way to do that is to carry your feedback culture over to other areas, not just technical proficiency.

Here are some other things you should think about when leveling up your engineers:

- ▸ How are their interactions with the team?

- ▸ Are they contributing by providing feedback on PRs on a consistent basis?

- ▸ Is there clarity in their communication?

- ▸ Are they focused at meetings?

- ▸ Do they show up on time?

- ▸ Do they meet their commitments or raise their hands when commitments are at risk?

I'm sure you and your management team could come up with other areas where you provide feedback. The point is to create expectations outside of coding excellence. When you don't, you're falling back into the stereotype of the asocial engineers who you leave alone because you don't understand them. Expect more out of them and you'll get more out of them. Well-rounded engineers are the ones who can drive your success by going beyond writing code. They'll be the ones helping the company think strategically and coming up with innovative products and features.

TROUBLESHOOTING COMMON PROBLEMS

As the story at the top of the chapter highlighted, you'll find that some engineers who came from other companies are at first struggling to adapt to a culture of constant feedback. It's understandable; they likely came from a company that had little meaningful feedback, but they soon come to love how much they grow from it. Here's how you should counsel them:

- Be patient with it. The fear of feedback will turn into the craving for feedback once you understand that it's all about mastering your craft.

- ▸ Go and look at everybody else's feedback, which is available and transparent. Everyone is getting it; it's not a negative pointed only at you.

Another common problem I see is a failure to commit to this kind of culture on a consistent basis. For example, you and your managers may be tempted to skip code reviews by saying, "We've got to get this feature out the door. Let's just roll with it as is."

Don't do it. You think you're achieving velocity, but what you're really doing is lying to yourself about how fast you're going and setting yourself up for huge headaches later.

Of course, going fast is good, but sloppy shortcuts aren't. The next chapter is all about how to gain velocity without getting careless.

Chapter 6

GAINING VELOCITY

For a startup, velocity is the difference between success and failure. It's that important. Go too slow, and you won't gain enough users or customers before you run out of money. Go too fast and... well, you can't go too fast as long as you do it the right way. Go fast and you'll grow and succeed.

As important as velocity is, it's a source of misunderstanding and frustration for many founders. I know from hundreds of conversations with them that they're not sure what to measure. And when they are measuring something, they're not even sure it's helping their company go faster.

I'd love to say these founders are worrying too much, but these are extremely valid concerns. Founders and the leadership team at many startups often do put too much emphasis on the wrong things.

This chapter and the next chapter are companions. This one will give you a big-picture view of velocity, and the next chapter will give you more specifics on how to go about optimizing your processes to go faster.

At the beginning, I recommend against trying to measure too much at once. Your company is likely not ready to analyze every leading indicator of velocity. Measuring in too many different directions adds confusion, when what's needed most at the beginning is simplicity and clarity. Also, if you're constantly measuring, it chips away at the time available for actual work to get done.

But that's almost a minor problem compared to another common failure. Many new companies fail to focus on the most important velocity metric of all. Instead, the company gets bogged down with an overemphasis on things like how many story points have been completed, how many features are being added, the number of bugs created versus number of bugs closed, and so on.

Let me quickly add that those metrics *do* have value, and they are key indicators in helping you understand how fast your company is going. But those are not *the* key metric, and to think that they are can mislead you into thinking you're achieving velocity when you're not.

What's the most important thing to measure when determining your company's velocity?

It's how fast you deliver value to your users.

Focus on this and you'll know if you're gaining true velocity. In the end, delivering value to your users at increasing speed makes the difference between a startup that fails and one that achieves remarkable success.

Founders are smart people, and they immediately grasp that delivering value to users as fast as possible is the fundamental metric. So why do they often lose sight of it?

One reason is that it's hard for most founders to understand what's going on "under the hood" when it comes to software engineering. It's just human nature to latch on to concrete metrics (like the number of story points completed) that feel like they're showing you how fast you're going.

It's especially tempting to focus too much on these metrics because they truly *are* one piece in the puzzle of velocity. They only become a problem when they're not related back to how fast you're delivering value to customers and users.

The larger issue with measuring how fast you're delivering value to your users is that it's as much art as science to figure out. When it comes to business and engineering, we like hard metrics that are completely clean. Story points? That's a number. Bugs created versus bugs closed? That's a clear ratio.

But what is the number for how fast you deliver value to your users? There isn't one—at least not in the hard-and-fast way that feels more comforting, more measurable. So many companies get lost in a maze of metrics that give the illusion of velocity without delivering value to users fast.

If it's not a specific number, is this metric just a feeling? No.

Here's one way you can think of it: how long does it take to go from the idea for a feature to your users being able to actually use that feature? This is not an exact science either—there will be valid reasons why some features take longer than others—but it is measurable and it does provide focus.

Another way this metric helps is that it serves as a guide for your daily ongoing discussions with your engineering leadership. Discussions about velocity will no doubt include the hard-and-fast numbers, but the fundamental orientation should always be along the lines of, "Okay, those metrics sound good, but when do we think we can put this in front of users?"

It's up to you as the founder to ingrain this into your company's culture. If your engineers know the first question out of your mouth on any feature is "How fast will we get this in front of users?" and "Is it possible to do less in order to deliver it faster?" do you think they might start focusing on it? Of course they will. Before you even bring it up, they'll be asking themselves what needs to happen for this to get in front of users and how can I speed it up.

The best possible thing you can do for velocity is to keep your whole team focused delivering value to your users.

What do you do if you're focusing on this and your team is still slow to get new things in front of users? That's when you turn to these more concrete metrics to look for clues as to why. But never lose the North Star of "how fast can we get this in front of users?"

There's another way to go faster: reduce the scope of whatever it is you're trying to release. Make a feature smaller. Don't solve for every edge case before releasing it. This applies even when you know your users will eventually need the extras you're cutting out. It's a slippery slope to want to have everything the users will eventually need done before you release anything new.

You can't be contradictory in the signals you send your team about shipping fast. Don't come down on people for shipping with less than optimal functionality. Most smart people are motivated to deliver the right solution with complete functionality the first time; they've been rewarded all their life for being right. But you need to send the message that it's okay to fail and learn fast.

Think of it this way: Instead of going for the eighty-yard touchdown, just figure out the play you need to get the next first down. Yes, eventually you're going to have to score that touchdown if you want to win, but for now, focus on getting ten yards. Reducing the scope is one of the best ways to get that first down.

THE DELICATE BALANCE OF TECHNICAL DEBT

Many startups take a ton of shortcuts when developing their first version in order to create a proof of concept, validate the market, and raise money. However, once you are at the point where you want to build a lasting company, it's time to be more calculated on the shortcuts you take.

Writing sloppy code is a bad long-term strategy, even if you experience a temporary short-term gain. What you're doing now is building the foundation that can make or break your company. It's a new mindset when you're at the point of scaling an engineering team.

There are times that you'll want to consciously build in bugs to be fixed later as a faster way to get something in front of customers. The best bugs on the planet are the ones you planned. The worst bugs are the ones you don't know about.

Any bugs you don't fix, any edge cases you don't solve, and any feature where you reduced the scope knowing you'll need to add more later, that all falls in the category of "technical debt."

Here's the crucial distinction between sloppy versus technical debt. If you're writing sloppy code with no real plan or conscious thought about why other than speed, that will end badly at some point. If you're consciously thinking, we know the mistakes we're committing, and we know what we're leaving undone but we're balancing it against speed, that's the right way. In that scenario, you're managing your technical debt.

Just like with credit card debt, there's a wrong way to manage it and there's a right way. The general outline of good management looks something like this: Put something in front of users knowing it has some technical debt built in. Validate that it's useful to your users. If it is, go back and "pay down" that technical debt, meaning fix the problems you left in there.

Here are some examples of technical debt:

▸ Edge cases: User scenarios that are less likely to happen, and you haven't yet created a solution to address it.

▸ Reusability: Code that is solving only the problem at hand but could be expanded to solve similar problems in the future.

▸ Unit tests: Code that tests code.

▸ Hardcoding: Data or parameters that can only be changed in the code.

▸ Faking actual functionality: Buttons or other features that seem to work but either have no functionality or makes the user think the system is doing something that it isn't.

▸ Scalability: The code might break with too many users.

THE FOUR LEVERS YOU CAN PULL
FOR GREATER VELOCITY

There are four ways to go faster than you are currently:

▸ Hire more engineers. We covered recruitment, but of course there's always an upper limit to how often you can pull on this lever. You have only so much cash.

▸ Invest in your people. We covered this concept in the last chapter, especially giving your engineers purpose, mastery, and autonomy. Happy engineers go faster.

▸ You can invest in technology. That's outside the scope of this book, but it's a lever a good CTO/ architect can help you with.

▸ You can improve and optimize your processes.

That last lever is the subject of the next chapter. Now that you have the overview of velocity, let's get down to some practical ways you can optimize your processes.

Chapter 7

PROCESS OPTIMIZATION

As a founder of a tech company, you likely have people recommending you head over to your local bookstore and grab the latest book on agile development, read it cover to cover, and then follow it to a tee.

My advice is a little different. Yes, I agree you should definitely understand agile development methods, and reading a book on it is a good use of your time.

But I've seen some companies waste too many resources and expend too much energy trying to implement it perfectly.

Instead of the concepts serving the company, the company is serving the concepts. The better way is to understand the ideas and then pick and choose those that will work for you and your team. It's more about adapting it to your company, rather than adopting it whole cloth.

My thinking is, I'd rather give you guideposts that you can fit around your company and integrate into your culture. This seems to me a better approach than forcing your organization to fit the mold of one particular philosophy or methodology.

These key six strategies below are ones I've used myself to get several companies to achieve massive velocity, and I've seen them work elsewhere, too. Any tech company that wants to optimize their processes to go faster should do the following:

- ► Learn about agile development and pull out what works.

- ► Keep the time your engineers spend in meetings to a minimum.

- ► Do efficient postmortems as part of a culture of continuous improvement.

- ▸ Define "done" for your engineers.

- ▸ Release features, stories, and so forth as often as you can.

- ▸ Use the concept of minimum viable product on an ongoing basis.

Let's take a closer look at all these and how you can implement each one.

Learn about Agile Development

There's a reason agile development has been around for a long time and why it's so widely adopted: it works. As stated above, my only problem with it is many people think you have to follow some version of it almost religiously.

You should use other resources to learn about agile development in more detail, but let me give you a brief outline of the two major frameworks within it.

One framework is Scrum, which incorporates several concepts including transparency, teamwork, and adaption. But at its heart is the idea of setting up "sprints" for your team.

The basic idea behind Scrum is quite simple: create short, time-bound goals and then focus the team on "sprinting" to accomplish them. Evaluate how you did, learn from it, set up the next sprint. Rinse and repeat, perpetually.

In practice, it looks like this:

- ▸ At the beginning of the week, you define what you think can be accomplished that week.

- ▸ At the end of the week, you compare what the team actually accomplished to what you set out to do.

- ▸ Analyze why it went like it did.

The other major framework in agile development is Kanban. This is more task based than deadline based. There's the list of things that the company needs done, and your business and product people are constantly reorganizing and reprioritizing in accordance with what is best for the business. As an engineer finishes one task, they can go to the list and grab the next thing that needs to be done. It cuts out a lot of process waste.

When you've recruited an excellent team, I lean toward Kanban. The artificial deadlines of Scrum become less necessary with a super-motivated team that has purpose and autonomy and cares about mastering craft. That's not to say Scrum won't work with a great team. They are both excellent frameworks and you should take some from each.

Cut Meeting Time

Long and/or unnecessary meetings are an endemic enemy of corporate culture. In a startup, they can delay projects and cost money by taking up precious time that could be used more effectively. Keeping an eye on meeting time is even more crucial during the early stages of your startup. Every minute that an engineer is in a meeting is a minute they are not writing code. If your goal is to get stuff in front of your users, your engineers need to be spending time writing code.

If your engineers are spending more than 20 percent of their time in meetings, you need to start asking questions and finding out why. What is being discussed and are there ways to solve the problem other than a long meeting? Can the meeting be done standing?

Also, maybe some engineers were invited to the meeting who don't really need to be there. For the most part, an engineer's life at your startup is going to be writing code, reviewing code, or attending a meeting. Cut down meetings, and the other part of the equation naturally goes up.

Of course, meetings are sometimes necessary, but focus your team on making them more efficient, with the lowest number of attendees possible to address the matter at hand.

Efficient Postmortems

One type of meeting I do recommend is postmortems. This fits nicely with the Scrum model but really works well in any methodology as long as there is a commitment to respectful honesty as you review any project.

One of the ways to set the correct tone is to choose the right question for a postmortem. The fundamental question should always be, "What could we have done better?" not "What went wrong?"

The best follow-up question to "What could we have done better?" is "How can we make sure we don't do that next time?" Don't settle for vague answers here. Look for ways to drill down and put in processes and checks that make sure it doesn't happen again.

Another crucial question is, "How could we have gone faster?" Look for tasks that took longer than normal and discuss ways to be more efficient next time. Your team leads should be taking note and look for opportunities to cut waste and increase velocity.

To keep things efficient, avoid attempting an exhaustive post-mortem. This is a common mistake. Whatever project you're analyzing, the team could probably identify at least twenty things you could've done better. That's way too many, and you'll waste tons of time.

Instead, identify three things that could've been done better, figure out how to fix them, and then implement those three solutions. It's always tempting to find more things that went wrong. Getting to a list of ten or more things that could be fixed would be easy.

Avoid the temptation. Trying to fix too many things fast at the same time is not only difficult, but it also muddies the picture of what actually improved, because there are too many variables to analyze. Focus on three things to fix at most. Always set this kind of short and fast tone for postmortems. Set a time limit and get it done, and then move on.

Efficient postmortems are at the heart of instilling continuous, incremental improvement into your culture. You will be amazed at how these contribute to exponential growth in velocity when you commit to them on a consistent basis.

Define "Done"

Your engineers need clear parameters to know when they have finished whatever coding task they're working on. This is a simple, commonsense principle, but it's not unusual to see startups who never clearly define what done means.

There's also occasionally the opposite problem where the definition of done is overdetermined and your engineers are wasting time wading through all the details. Make the definition of done as simple as possible while still having it be clear.

The three areas that need to be defined are:

▸ Business or functional requirements: A list of acceptance criteria that meets the need of the problem.

▸ Quality: Parameters for what's acceptable in edge cases, unit tests, and other technical debt

▸ Nonfunctional requirements (NFRs): Address scalability, security, usability, compliance, and all other relevant areas.

Quality and NFR lists are usually smaller when you start your company. They will naturally grow as your company becomes more mature and needs to pay closer attention to things like compliance and security. The key point is that your company should be on the same page about what done looks like and then to revisit it as needed. You may find that some situations will require different definitions than others based on what you are trying to accomplish.

You can work out the nuances of defining done with your engineering leadership, but the overarching point is this: as an organization, your startup needs to have a clear definition of done.

Fast Release of Features, Stories, Versions, Etc.

The previous chapter explained that velocity should be defined as how fast you are getting value in front of your users or customers. One of the things that I said to measure was how fast something went from idea to getting in front of your user.

There's another way to measure this, too. You can look at the rate at which you are releasing specific things to your users overall.

Let's take features, for example. How often are you putting new ones in front of your users? Measure that. Is it once every two months? Could you be doing it once a month? Could you get it down to once a week? How about daily, eventually?

You may not be releasing major features every day, but can you get to the point where you are shipping something daily.

The answer for the quantity of new things you can put in front of your users will grow over time if you're doing it right. Daily or even weekly is probably not realistic for a startup in the process of adding its first round of engineers. But you might be surprised by how fast you can get there if you focus on this metric.

How to Think about the Concept of Minimum Viable Product

I'm sure I don't need to educate you on the idea of delivering an MVP to your users. It's a bedrock principle to put an MVP in front of your beta testers and early adopters and get feedback so you don't waste tons of time and money doing things they don't care about. Obviously, this is a crucial concept for any startup.

But I've noticed that many tech companies, both established and startups, don't extend this concept nearly far enough. The common idea most have of an MVP is that it applies to only the first version of your product. Once your product is adapted more widely, there's a thought that every subsequent added feature, story, or version needs to be much more than minimally viable.

That's bad for velocity. You should treat everything you release as an MVP. The temptation to make everything perfect is understandable when your base of users or customers is much wider, but it's a mistake because it kills velocity without much benefit.

Think of it this way. If you revert to going too slow, that's a hard problem to fix and it will take time to do it. If you're going too fast, your users will tell you that. It's a great problem to have and one that's much easier to fix. You can do things like release to beta testers to solve that problem without killing the momentum of your team.

Remember, everything you release—features, versions, stories—should be an MVP!

YOU CAN *FEEL* VELOCITY

You want to know one of the best feelings you can have in your startup? When you know in your bones that your team has achieved amazing velocity. You'll of course keep using metrics and analyzing them. But you won't need those measurements to know that you have an amazing team that's moving at the speed of astonishing success.

You'll see it in how your team interacts with one another. The postmortems will become razor sharp, everyone will be on the same page about what "done" means, and pointless meetings disappear. Focus on the six keys to optimizing your processes and it will build slowly at first, but then you'll feel that velocity.

There's really only one area I can think of where you don't want too much velocity, and that's handing out titles.

Chapter 8

DO TITLES MATTER?

You're at a startup, so titles don't matter, right?

Getting the work done matters. It's you and your surrounding core against the world. Titles are the kind of thing big organizations obsess about, and they may be a necessary evil at that point, but at startups we're focused on getting important stuff done and titles are an afterthought at most.

Well, kind of.

If your startup team recognizes that titles aren't nearly as important as getting things done, that's a good thing. It's a healthy attitude to cultivate in the early stages of a company. There will be a lot of opportunity for career path growth and latching on to titles later, but for now, it's about pulling everyone's energy together and making sure the company survives to become successful. Without that, everyone will be looking for titles somewhere else anyway.

If that's what you mean by "titles don't matter," then perfect.

But from another perspective, titles very much do matter from day one. They will be important to the future health of your organization, and if you give them away too soon, expect pain later. The best way to illustrate this is by showing you what can go wrong.

I've seen many cases where the CTO title is given to the first engineer the founder had on board. If the person is qualified to be a CTO, this is ideal. But there are many cases where a wonderful software engineer is not also the right person to be CTO.

The CTO is the person who is the guardian of your technology. They'll be guiding you on programming languages, architecture, vendors, and other foundational technical decisions, ones that stay with your organization forever.

It's an absolutely crucial role that requires shrewd, tough decisions. You're going to have a lot of smart engineers on board with multiple good ideas on technology choices. You need one go-to person who can make the final decision. The wrong person in that role can be fatal to your company.

It's understandable that a founder would give the CTO title to the first engineer they hired. That person often seems qualified to a founder because that engineer knows more about your code than anyone else. And again, sometimes they're genuinely qualified to be CTO. But don't just give them the title and position because they were hired first. As a founder, you likely don't have much of a technical background. Your CTO is your protector and translator in the world of technology. Be careful about whom you give that title to!

Another thing that I've seen are companies that have five engineers on board, but somehow, they already have two Directors of Engineering. In cases like these, usually the founder has lured them away from another company where they already had a director title.

These engineers understand that the bulk of their job early on will be coding, and they are okay with those tasks—that's great. They just don't want to come on board without the title they previously held—that's not so good. Use other incentives than a title to get them on board. If they're excited about your mission, a lesser title is something you can usually overcome.

Along these same lines, I've observed many startups hand out VP, Director, and Architect titles too early. Later, when the organization becomes more mature, I've seen these improper titles come back to haunt founders because it's too difficult to demote people or reset titles.

You might look at the above examples and agree handing out gratuitous titles to keep people happy is not ideal. But you also might be thinking something like, "Those founders were doing what they had to do to get people on board. At startups, you do what you have to do and figure it out later."

I strongly recommend against the "figure it out later" attitude when it comes to titles. Resolving a difference by kicking it down the road leads to problems and pain later. Founders who hand out titles willy-nilly don't always understand how much pain. As you grow, you're going to need the people with those titles to actually be able to do the jobs that correlate with them. You can't have two CTOs, so you better not have given the title away to the wrong person.

You'll also be creating havoc with career paths with misnamed positions. If you've given away several Director of Engineers titles, promoting others to that position later is going to create confusion. The career paths in your organization are going to be a bewildering muddle. That extinguishes motivation because people will have trouble seeing their way forward.

Then you'll be in a tough spot, with three options, and none of them good. You could keep fumbling through with the wrong people in the wrong positions (you already know that's a terrible idea). You could add another position with the same title, but that is embarrassing for all parties and a recipe for more chaos. Or you can have that exceptionally difficult conversation with someone about giving up their title.

If you find yourself in this position, it's probably the last solution that is the lesser of the evils, but wow, is it hard to have that discussion. You might even lose someone who is exceptionally valuable because of it.

It's far better to keep yourself out of this fix in the first place by being smart about titles right from the beginning.

You should also consider this: if the only way someone will come to work for you is if you give them a certain title, they may not be the best hire. It's better to spend time with them and explain that you believe that appropriate titling is important. Also explain to them that your organization hasn't built out career levels yet and you want to create a flat organization to start. As discussed in the hiring chapter, you want engineers who align with your mission, who are excited about the work you do and about the growth and learning opportunity the job provides.

THE BETTER WAY FOR STARTUPS

The focus so far has been on things that go wrong. Let's reverse it and look at the ideal for startups.

First, let me give you a rule you should always follow: only give a person a title if it describes their current tasks accurately. Don't speculate about where they will be when the organization grows. Don't give them a label as an incentive to get on board. Just be accurate. (Accuracy means matching it to how most other tech companies in the world describe the position's duties.)

The second rule I'd give you is less is more. Startups are often so fluid at the beginning that someone may be doing *some* of the role of a particular title. But stick with their current title until that higher position becomes fully defined as theirs and they prove they can do all the associated duties.

The final guideline I will give you is to be particularly slow to give out titles with any of the following in them: Chief, Architect, VP, and Director. Those will be important to the future health of your organization.

Early on, titles that reflect roles rather than position can be more effective. Sometimes a founder will come to me and say, "I have a mobile developer and a back-end developer and now I need to hire more people." Perfect. No set titles, just roles.

As you add more people, or if you already have a slightly larger core around you, the ideal is to have software engineers, senior software engineers, and engineering leads, and probably no more. (Unless you get lucky and have a qualified CTO or VP already.)

When you stick with these titles, you have maximum flexibility for future growth. I like the term "engineering lead" over "manager" because your engineering lead is likely wearing a lot of hats right now. The biggest hat should be writing code. Engineering lead implies someone writing code who is also in a leadership role. Manager starts to make it sound like they are exclusively managing others, not writing code.

Eventually, you'll have engineering managers who are coding less and managing more, but that won't be a concern until you grow to more than twenty engineers.

You might not even need a lead title at first. I've seen some flat organizations where everyone at the beginning is a direct report to the founder. That's okay, too, because you're being prudent from the start and can make the right adjustments fast.

CAREER PATHS AND MOTIVATION

A big part of the reason you want to be sparing with titles early on is you're going to need them for motivation later. Clearly defined career paths with accurate titles will be another source of incentive for your engineers. As positions are created through organic growth, it allows those within the organization to fill them based on merit and natural talents.

The engineers on your team will aspire to newly created positions like senior manager, director, VP, and senior VP and will work for it. If those titles were given out easily and arbitrarily in the early days, there will be no sense to the career paths within your company. At best, engineers will lack motivation for growth; at worst, they'll be demotivated, which means they may begin looking for a job elsewhere.

I want to end this chapter with some good news. I think some founders overestimate how much titles matter to the engineers who come on board early. The kind of person willing to be one of the first ten to twenty at a startup doesn't do it for titles or for an immediately clear career path. There are plenty of established companies they could work for if they wanted that.

They're doing it for the excitement and because they believe in the mission. "Here we are creating a company from nothing and

I'm part of that." It's often the lack of firm titles at the beginning that is part of what they love. If someone requests one, don't assume that you have to say yes to get them hired or to keep them. It's often a very minor motivation overall.

So we've come full circle. Titles don't matter at a startup *unless you give them away too fast and too easily.*

Now that you've spent a lot of time recruiting, motivating, and optimizing a high-performance engineering team, how do you keep that team healthy?

Chapter 9

KEEPING THE TEAM HEALTHY

No matter how good your culture, there will be times when previously excellent engineers begin to show a drop in performance. What's interesting is that some company cultures accept this, often out of fear of the challenge of hiring more engineers.

Within companies that accept mediocrity over time, only a huge performance deficit will get addressed. If you want to keep your team healthy and performing at a high level, you'll need a different approach.

The foundation is understanding this simple principle: engineers should always get better over time. More experience and company knowledge should always fuel an engineer's growth.

Unfortunately, we know that not every engineer will get better, and some will get worse. The key in these situations is to figure out the cause. Because there has to be a reason. Good engineers don't deteriorate for no reason.

Here are the symptoms you'll notice as performance drops:

- Everything starts taking them a lot longer.

- They'll try to find ways to pass off work to teammates.

- Previously animated in meetings and discussions, they now barely participate.

- They make commitments and don't meet them, and don't even raise their hand early to say they won't make the deadline.

- Excuse making.

- Missing work too often.

Some of the symptoms will be more subtle than others, so it's something to keep a close eye on. Once you recognize the signs, you need to correctly diagnose the problem.

IS THE PROBLEM
ORGANIZATIONAL OR INDIVIDUAL?

The first thing you need to ask yourself is whether the problem is possibly with the organization and not the individual. To find out, I recommend returning to the ideas of autonomy, purpose, and mastery.

As you grow, your organization will have a natural tendency to take autonomy away from your engineers. You'll begin adding positions like product manager, for instance. An engineer who in the early days had a lot of say in the products that were developed will now have less as the product manager takes on many of those decisions.

This is a natural by-product of company growth, and the problem can't be completely eliminated, but it can be managed. For example, engineers should still have input on product direction and those ideas should be welcomed. If they're not, you need to figure out why.

This is where you should remember the difference between engineers as artists versus engineers as factory workers. If all your organization does is go to engineers with solutions instead of problems, that's a clear sign that you are creating too much of a factory worker mentality, and you need to put engineers in a place where they can also create.

Even more importantly, engineers can be given a lot of autonomy on *how* they build that new product or feature (assuming they've earned it and can handle it). Oftentimes as startups grow, they add so many managers that the autonomy of perfectly capable engineers becomes too limited.

That is not a natural by-product of growth; it's just poor management. If this is a problem, it will begin impacting more than one or two engineers and can be a cancer on your organization. Dig in and find out where the right balance is between management oversight and autonomy.

HAVE THEY
LOST THEIR PURPOSE?

The next thing to look at is purpose. Has your struggling engineer lost that? It could be what I call problem fatigue. It means they've been working on the same issue for five years, the goal each time being to make the same thing faster, better, and so on.

Progress on these issues can be satisfying, but at a certain point, it's still the same problem. You can tie the problem and goals back to the mission as much as you want, but eventually, solving the same problem over and over no longer inspires. (I should add that some people love working on the same problem and never get bored with it. Value them!)

When possible, look for ways to move the ones who are fatigued to a new problem. If it reignites their purpose, the performance will go back to a high level. Keep an eye on it and rotate them when you can.

ARE YOU GIVING THEM CHANCES TO MASTER THEIR CRAFT?

Finally, look at mastery. Are you still giving them ample opportunities to master their craft? It's easy to forget what it feels like from the engineer's perspective to be taken for granted as the company grows. Maybe they have a certain area nailed down. That's nice for the company, but over time, that can become a rut where they don't feel challenged and they never get to work on hard things.

There's usually a solution here. If you're an organization that has a competitive edge and wants to win, which I assume you are, there's almost always bigger, harder problems you should be tackling. Can you give your engineers enough of those to keep them growing in their craft? Look for ways to do that.

Of course, it's always a balance. You can't take on twenty new initiatives just to keep your engineers from getting restless. Mundane work still needs to be done. But if you want to keep good engineers, never stop looking for ways to balance the commonplace with challenges that spur mastery.

Once you've explored the organization for shortcomings, you correct for what you can and the rest you manage. But there comes a point at which for certain engineers, no reasonable adjustments work and performance stays low.

Now it's time for a deeper conversation.

To prepare for the meeting, come with stats and anecdotes about their days of high performance. Paint a picture of how they were then. Now compare it to how they are measuring up to that standard now. If you do it with fairness and honesty, it will be an inescapable conclusion for them that this isn't working.

You might be surprised, but about 90 percent of the time, this goes well. You've laid out the evidence before them, and they acknowledge it. Sometimes they are only fully admitting to it themselves when you show it to them.

It's also a good idea to point out to them that this is a normal human thing to happen when internal passion for something dries up. "Nobody who wakes up every day dreading their work can keep at it. There's no amount of carrot or stick that will make it work."

Dive into what they think the problem is. If it turns out to be a solvable problem related to purpose, autonomy, or mastery, experiment with making adjustments and monitor how it goes.

Of course, sometimes you discover that the issues are personal. Maybe it's a relatively simple need for some extra time off for a vacation or dealing with a family situation. When it's a personal problem, look for ways to help where possible.

After you've explored the reasons for the drop in performance, you'll sometimes not find anything you can help with, whether it's work related or personal. That's when the discussion will turn toward whether they'd be happier elsewhere. Your role here goes back to being a mentor for their career. "Let me help you find out what you do want and what will make you happy."

Often, the discussion will end with the engineer saying they want to think about it and coming in the next day telling you they agree it's time for them to get a fresh start. If you have their best interests at heart, you'll agree with them and even help them with the transition.

I've found one of the most fulfilling parts of my career is seeing engineers who used to work for me grow into huge roles at other companies. This applies whether they left under the kind of circumstances we're discussing now or otherwise. It all goes back to caring about people whether they're useful to you or not.

One subset of the problem that I'm talking about here is common enough that I want to address it. That's the engineer who can stay interested in a job for about nine months (give or take three months) and then they have to move. As I said back in the recruiting section, you want to screen anyone out who seems to have this problem of extreme restlessness. But if you get someone like this on board, it probably is not going to last and there's little you can do about it usually.

The one thing you can try if you grow big enough: constantly rotating them to new challenges (assuming they're valuable in each role). Not every company has the luxury to play musical chairs, but if you do, you can try this as a solution.

The one thing I don't recommend is to throw more money at them. More money will not solve an "I get bored easily" problem.

BLINDSIDED

There's another kind of parting with an engineer, and it can hurt. That's when you are totally surprised when someone gives you notice.

When you've invested in connections and feel a lot of caring toward the person, it can feel like a blindside attack. Hard as it can be, don't take it personally.

For the majority of engineers who leave my team, I know it's coming. It's either a scenario like we outlined above, or the engineer feels comfortable enough with me to discuss that they're ready for their next opportunity and have begun to look elsewhere. But there is a subset who just won't say anything despite the relationship you've built.

Job security is a powerful motivator, and some people are worried that if you say, "I'm exploring other opportunities" the reaction they'll get is, "You're fired." No matter how much you encourage honest dialogue, they surprise you with two weeks' notice.

When this happens, talk to them and find out why they want to leave. You'll learn from it if nothing else. And please don't stop them with guilt trips. For one thing, that's just wrong. For another, if it works, they'll resent it and will likely leave soon anyway.

If they say the reason they're leaving is for more money, should you match the offer (assuming you think the extra money matches their value)? Most of the time, no, and that's because it's rarely just money. If they say that's the motivation, dig deeper and find out for sure.

I've found sometimes they're happily working for you, and somebody pings them about a job. They start listening, they hear a higher number, and they say to themselves, "Maybe I should. It is more money." If I talk with them long enough and it really does seem to be only the money and I think they are worth it, I will match.

I don't do it often, and even when I have, I find it doesn't always work. Many times, they leave for another offer soon after. So match offers very sparingly.

AS THE COMPANY GROWS, SO DOES THE POTENTIAL FOR POOR PERFORMANCE

One question I get about poor performance is this: does it happen more as companies get bigger?

It can. You may have heard the analogy about the difference between startups and mature companies: are you in the jungle or are you on a highway? The young days of a startup are the jungle. You've got the adrenaline of the unknown. It's hard to see which direction to go. Is that tiger about to eat you? After your company reaches a certain maturity, you're on the highway. You've got your GPS, you know exactly how many miles to reach your next destination, you use your turn signals, and all of that.

The first one is scary but thrilling. The second is comforting and can lack excitement. People can get bored with the comfort, and performance drops.

However, never let the fact that you get closer to the highway be an excuse to allow low performance. For one thing, it's far from the only reason performance drops. I've seen people start to lose steam from day one and keep descending. And if you keep looking for ways to give engineers purpose, autonomy, and mastery, most of them will keep performing at a high level, even on the highway.

Always remember as the founder that a key part of your role is to keep your team motivated at all levels. Keep them pointed toward the North Star of the company mission. Create clarity on what company success looks like. Continuously celebrate wins, contributions, and improvements. If you don't show you care about this, no one will.

Leaving you with a final thought on keeping your team healthy, remember what's at stake. It's not just that the underperforming engineer is not pulling his own weight. That engineer is also infecting your team, pushing work to others, often being negative, and failing to give good feedback. Don't let things fester and spread. Address it.

Chapter 10

SHOW YOU CARE OR NO ONE ELSE WILL

I remember one time a founder shared with me his deep frustration that his lead engineer was leaving for a planned vacation, even though the company had just scored a round of funding and needed to make a special push to get more done.

The founder felt the person should've sacrificed their vacation because of the changed circumstances. It had him questioning whether this person was really committed to the organization.

I challenged him to look at it differently. Maybe the problem was the reverse, that he expected commitment to go only one way. After all, this lead engineer had her own life, and the vacation was no doubt important to her. Wasn't it the right thing to commit to her by letting her go on a vacation without resentment?

The founder decided to take this advice and benefited greatly by retaining a very valuable team member. Eventually, that lead engineer became an executive at this company, which is successful to this day.

If you're a founder, you have to accept that no one is going to sacrifice as much as you do for your company. That's how it should be, right? If someone cares more about your company than you do, something's wrong.

Yet, I often talk with founders who have a sense of disappointment, sometimes even anger, that the people who work for them don't put in as much time and make as many sacrifices as they do.

If that's you, you need to embrace the following statement and its consequences:

No one cares about your company as much as you do. And that's not going to change. Period.

Then when an engineer leaves at 5:00 p.m. to take a spouse out to dinner when there's still a problem that needs to be addressed, you won't be surprised or upset. People are going to sometimes put other things ahead of work, even some things you think should be sacrificed. Be okay with it.

There is something that you can do to get more sacrifice, dedication, and commitment from others, however. The secret to getting people to commit is by giving, not taking.

Reminding people constantly that "we're building something together and we'll all win when it succeeds" is about taking, not giving. It's good to remind your team of that "we're building something" message, but that only goes so far when it's all you do.

If you want commitment from them, you need more than inspirational team speeches. The better way is to commit to them as people. Throughout this book, from the very first section on recruiting onward, I've talked about caring about your engineers as individuals and having their best interests at heart.

It might sound kind of touchy-feely to you, but it works. And I don't think it's touchy-feely; it's good sense. It shouldn't be done craftily or anything like that. It's as simple and as hard as connecting with people and finding out what they want for their careers and their lives (remember back to Chapter 4 about getting to know your engineers).

You won't get this knowledge from impersonal surveys with standard questions. "What would you like to be working on? What are your hobbies?" No, that's not the way. It's spending time with them and proving deep commitment by your actions.

Whenever I talk about this with founders, I fear some will think this is a calculated technique for manipulating people into working harder. I've no doubt that faking commitment to people could work on a limited number of people for a limited amount of time.

Most people see through that, though. You're working closely with them, and after a time, the truth will come out. Not to mention, I can't understand why anyone would want to lead a fake life when it's so fulfilling and satisfying to help people get what they want out of their careers and their lives.

Another way to look at this is that getting the most out of people is a side effect of doing the right thing. The goal is not to squeeze as much work out of your team as you can. The goal is to care about them and then have the side effect be their hard work and dedication to the company.

It is okay to be honest that it's a two-way street, but not if you put it in terms of "I've done this for you; now you should do this for me." Instead, it's "I commit to you and your success, and I'm hoping you feel the same about me and this company." All the difference in the world.

I've seen some founders work in the opposite direction, and it's not pretty. One particularly negative thing I've witnessed is those who lash out at employees when they think they have prioritized other things over work. That's an absolutely terrible idea. You should be setting the example of a healthy perspective on work and showing it in your actions.

Instead of being nasty when you think somebody is not working as many hours as you, compliment them for making a decision to do something with their family. You might be surprised that the next time something really needs to be done, that person is the one staying.

BANISH THE TERM
"WORK-LIFE BALANCE"

You might think I'm driving toward the concept of work-life balance. Nope. There's something at the heart of that phrase that is completely misguided.

Set aside that the term is overused and often uttered by people who could care less about the life side of your equation. But even people who have good intentions don't realize the message it's sending.

What it implies is this: work is so grinding and stressful that you need to find some relief in whatever time off and recreation you can squeeze out around it. The term "balance" implies that one side of the equation is relaxing and good (life) and the other side is tense and bad (work). That's the wrong way to look at it.

Work should be a *part* of your life and an extremely positive one. It shouldn't have to be balanced out. Do you need to rebalance after spending a day at the park with your kids or hanging out with friends? Why should you need relief from a job you love?

You spend more than 50 percent of your waking hours on weekdays at work. When it is an integral part of a healthy life instead of something to run from, the whole idea of clock watching or struggling to make it to the weekend evaporates.

A quick note that I have found some of the most dedicated and productive people are regimented about their time and will leave when "the whistle blows," so to speak. Never hold it against them. Let them be who they are if they're working hard and being productive.

I hope you didn't misunderstand any of the above to imply I want you to tell people to not worry about time off or having weekends or going home at a reasonable hour. I advocate for all that. In fact, you should be a leader by example in that area. Just as work is not drudgery, neither should it be the only source of happiness in a person's life.

If the demands of your company are so high on you at the beginning (which admittedly can be the case for a founder), you need to make clear to everyone that you're not the standard in this area. If you're putting in a crazy number of hours and you don't tell people explicitly that you don't expect them to do the same, you'll create a lot of unspoken anxiety.

The anxiety will come from your people who will feel guilty about whether they should be there all the time, too. They'll feel anxiety because they're not sure of the expectations in this area and will feel like they're falling short.

Anxiety and guilt create burnout. Burnout leads to quitting. Losing too many good people can kill your company. Don't make your own schedule the standard, especially if it's the early days and you're working like a demon.

I want to bring all the themes of this chapter together and talk about the magic it can make happen.

- Show you care by committing to the engineers, and do it by actions, not just words.

- Make work a part of a great life for your employees, instead of drudgery that needs to be balanced out by other activities.

- Make it clear that you care about the impact someone makes, not the number of hours they work.

Now here's the magic. If you do all these things with sincerity, people will want to run through walls for you. Engineers are smart people. There will be times when they know something needs to be done before they leave, or it will cause a problem for the company. If you do all of the above, 99 percent of the time, most engineers will make the choice to stay late and make that crucial thing happen.

In my experience, when you treat your team like adults, they act like adults. It's important to remember that people have different work styles. Most will know when they have to push themselves to get something done versus when they need a break. Trust them to know when they need a small break to take a walk in the middle of the day and when they're having a day when their brain is fried and they need to sign off early. Recall the story at the top of this chapter, remembering that people have needs outside of work and sometimes need a special vacation or time off to meet a personal commitment.

Every once in a while, you will have someone who hasn't ever been treated like an adult in a past job or is new to the workforce. In this case, work closely with them to help them understand what it means to perform at the highest level and to listen to their gut when it comes to pushing the envelope on burning the midnight oil or taking the breaks needed in order to have a healthy life.

This leads into one more vital point. No matter how smart a person you are and no matter how good of a leader, every day in your company, other people are making decisions for you. Most times, they are small decisions that accumulate every day for a big impact later. This is especially true of your engineers. Some of them may not even think of it as making decisions, but every time they figure out how to make something go faster or work better, they are making decisions for the company.

Your company will eventually live and die by the cumulative effect of all those decisions made by all the key people working for you.

Do you want them resentful or fearful of you because you seem like you always want just a little more from them but never seem committed to helping them achieve what they care about?

Or do you want people who know how committed you are to them and are ready to run through that wall for you?

There's no more important advice in this book than this: show you care.

CONCLUSION

Maybe the information in this book has fully converted you to its concepts for building a high-performing software engineering team. I'd love that, obviously.

However, even if that's the case, I'm guessing there's something nagging at you.

How in the world, with everything else you have to do to build a company, are you supposed to accomplish all this?

Building an entire company is certainly a full-time job and then some. But clearly, recruiting, motivating, and optimizing a superb engineering team is also full-time work and then some.

Here's my recommendation. Do as much of this as you can at the beginning yourself. You'll learn a ton, you'll bond your engineers to you, and you'll be able to understand what parts of it you're good at and what parts not so much.

However, there will come a time, maybe it's now, maybe it's a few months or a year away, when handling it will no longer be workable for you or in the best interest of your company.

Then it's time to look for a VP of Engineering if you have the resources to do it. This book will help you with that search, too. I find that founders who want to immediately hire a VP of Engineering don't know what questions to ask them. You could find the most qualified VP in the world, but if they're not aligned with the way you want to build your company, it's still a bad hire.

By understanding the concepts in this book and trying them out a bit on your own, you'll be in a much better position to find the right VP of Engineering for your goals. I'm sometimes asked if the CTO could or should handle these duties instead of adding a VP.

CTO and VP of Engineering are titles that get a little confusing in the tech world because companies often use them to refer to different things.

My strong opinion is that the CTO is in charge of technology. The title is Chief Technology Officer, so I think that makes sense. As I mentioned in Chapter 8, the CTO is the guardian of your technology and will decide crucial things like the language you use and the architecture you implement.

I believe that healthy organizations will also have a separate person focused on building the team, someone who concentrates on the people side of it. That person's title should in most cases be VP of Engineering.

Companies that have one person trying to fulfill both CTO and VP of Engineering roles will probably struggle mightily. Being in charge of day-to-day technical decisions and maintaining the health of the company's code base is a full-time job for the CTO. If you then add the responsibility of building a high-performance engineering team, the burden becomes overwhelming.

As I said above, it's my strong belief that the proper titles for these roles are CTO and VP of Engineering. However, whether you take my advice on specific titles or not, you definitely need to make sure roles are clearly defined.

Some companies reverse the titles and put the CTO in charge of building the organization. In that case, the CTO usually has lots of people support and does not have to be as connected to the health of the code base. Whatever you do, make sure you have someone guarding your code, technology, frameworks, and infrastructure. These are the lifeblood of your organization and you can't risk your company imploding from a major technical issue. Having someone who has a deep understanding of your technical assets who sits at your executive table is critical, whether you call that person the CTO or not.

If you recall back to Chapter 6 about velocity, there are four levers that can increase it: recruiting, investing in your engineers, optimizing your processes, and making the best technology choices. My recommendation is to have a VP of Engineering focus on recruiting and investing in your engineers. The CTO should own the technology choices. And the two should work together as partners in optimizing your processes.

Even if you had someone who was a genius at both technology choices and building a high-performing engineering team, there aren't enough hours in the day for someone to do both at a growing startup.

Of course, when you do make the decision to find a VP of Engineering, you're putting a layer between you and the engineering team. The key is to make sure that the new VP is aligned with your wishes for building your company.

You're going to want to know things like:

- "Talk to me about your hiring philosophy."

- "How do you see your role in growing people's careers?"

- "How do you go about optimizing processes?"

- ▸ "What do you believe it takes to create high-quality software?"

- ▸ "What do you measure when it comes to understanding the health of your organization?"

- ▸ "What do you enjoy most as an engineering leader?"

I hope this book has sparked you to think about how *you* want to build your engineering team, and now you'll know the answers you want to hear from any prospective VP of Engineering.

I'll leave you with this thought. Everything changes as you grow. If your startup grows to terrific success, you'll need to use recruiters, you'll need to hire to fill positions all over your hierarchy, and so on. Everything will have to scale, including hiring, motivating, and optimizing.

But the core concepts shouldn't change. The culture of feedback, the bedrock principles of purpose, autonomy, and mastery, and genuinely caring about the people who work for you all stay the same. It's up to you to make sure your company never loses that culture that made it successful in the first place.

ABOUT THE AUTHOR

DAVID DETTMER has been a leader in product engineering for startups for more than twenty years. He's built several successful software engineering teams from scratch and has also consulted with hundreds of startup founders on how to hire, build, and optimize high-performing engineering teams. David lives in Austin, Texas with his family and has raised his own engineer, his daughter.